Arne Micheels

Late Miocene Climate Modelling with ECHAM4/ML

Arne Micheels

Late Miocene Climate Modelling with ECHAM4/ML

The Effects of the Palaeovegetation on the Tortonian Climate

Südwestdeutscher Verlag für Hochschulschriften

Impressum/Imprint (nur für Deutschland/ only for Germany)
Bibliografische Information der Deutschen Nationalbibliothek: Die Deutsche Nationalbibliothek verzeichnet diese Publikation in der Deutschen Nationalbibliografie; detaillierte bibliografische Daten sind im Internet über http://dnb.d-nb.de abrufbar.
Alle in diesem Buch genannten Marken und Produktnamen unterliegen warenzeichen-, marken- oder patentrechtlichem Schutz bzw. sind Warenzeichen oder eingetragene Warenzeichen der jeweiligen Inhaber. Die Wiedergabe von Marken, Produktnamen, Gebrauchsnamen, Handelsnamen, Warenbezeichnungen u.s.w. in diesem Werk berechtigt auch ohne besondere Kennzeichnung nicht zu der Annahme, dass solche Namen im Sinne der Warenzeichen- und Markenschutzgesetzgebung als frei zu betrachten wären und daher von jedermann benutzt werden dürften.

Verlag: Südwestdeutscher Verlag für Hochschulschriften Aktiengesellschaft & Co. KG
Dudweiler Landstr. 99, 66123 Saarbrücken, Deutschland
Telefon +49 681 37 20 271-1, Telefax +49 681 37 20 271-0, Email: info@svh-verlag.de
Zugl.: Tübingen, Eberhard Karls Universität, Dissertation, 2003. Revised version 2009.

Herstellung in Deutschland:
Schaltungsdienst Lange o.H.G., Berlin
Books on Demand GmbH, Norderstedt
Reha GmbH, Saarbrücken
Amazon Distribution GmbH, Leipzig
ISBN: 978-3-8381-0758-5

Imprint (only for USA, GB)
Bibliographic information published by the Deutsche Nationalbibliothek: The Deutsche Nationalbibliothek lists this publication in the Deutsche Nationalbibliografie; detailed bibliographic data are available in the Internet at http://dnb.d-nb.de.
Any brand names and product names mentioned in this book are subject to trademark, brand or patent protection and are trademarks or registered trademarks of their respective holders. The use of brand names, product names, common names, trade names, product descriptions etc. even without a particular marking in this works is in no way to be construed to mean that such names may be regarded as unrestricted in respect of trademark and brand protection legislation and could thus be used by anyone.

Publisher:
Südwestdeutscher Verlag für Hochschulschriften Aktiengesellschaft & Co. KG
Dudweiler Landstr. 99, 66123 Saarbrücken, Germany
Phone +49 681 37 20 271-1, Fax +49 681 37 20 271-0, Email: info@svh-verlag.de

Copyright © 2009 by the author and Südwestdeutscher Verlag für Hochschulschriften Aktiengesellschaft & Co. KG and licensors
All rights reserved. Saarbrücken 2009

Printed in the U.S.A.
Printed in the U.K. by (see last page)
ISBN: 978-3-8381-0758-5

Table of contents

ABSTRACT	v
1 INTRODUCTION	**1**
2 TORTONIAN REFERENCE SIMULATIONS WITH ECHAM4/ML	**5**
2.1 The model ECHAM4/ML	*5*
2.2 The Standard Tortonian run	*6*
2.2.1 The model setup of the Standard Tortonian run	6
2.2.2 Results of the Standard Tortonian run	8
2.2.3 Verification of the Standard Tortonian run	11
2.3 The $2\times CO_2$ Tortonian run	*12*
2.3.1 The model setup of the $2\times CO_2$ Tortonian run	12
2.3.2 Results of the $2\times CO_2$ Tortonian run	12
2.3.3 The comparison of model results and quantitative terrestrial proxy data	16
2.3.4 Verification of the $2\times CO_2$ Tortonian run	19
3 THE RECONSTRUCTION OF THE TORTONIAN VEGETATION	**22**
3.1 The method	*22*
3.2 The resulting Tortonian vegetation	*26*
4 THE PALVEG TORTONIAN RUN WITH ECHAM4/ML	**28**
4.1 The model setup	*28*
4.2 Model results of the PalVeg Tortonian run as compared to the Standard Tortonian run	*28*
4.2.1 The global average temperature, precipitation and sea ice	30
4.2.2 The zonal average temperature	32
4.2.3 The regional temperature, precipitation and evapotranspiration patterns	34
4.2.4 The large-scale atmospheric circulation	38
4.2.5 Regional atmospheric circulation patterns	49

4.3 The PalVeg Tortonian run compared to the Recent Control run   55
4.3.1 The regional temperature patterns   56
4.3.2 The global average precipitation and evapotranspiration   57
4.3.3 The zonal average precipitation and evapotranspiration patterns   58
4.3.4 The regional precipitation and evapotranspiration patterns   59
4.4 Discussion   62
4.4.1 Weak points of the model and of the setup of the PalVeg Tortonian run   62
4.4.2 The comparison of the PalVeg Tortonian run with other model results   64

5 VALIDATION OF MODEL RESULTS WITH PROXY DATA   70
5.1 Methods and data   70
5.2 The quantitative comparison   71
5.3 The qualtitative comparison   74
5.4 Discussion and summary   77

6 VEGETATION MODELLING WITH CARAIB   81
6.1 The CARAIB model and its setup for the Tortonian   81
6.2 Results of CARAIB simulations   82
6.2.1 The simulated vegetation   82
6.2.2 The carbon cycle   87
6.3 Discussion   88

7 SUMMARY AND CONCLUSIONS   91

Acknowledgements   95
References   96
Appendix A - The PalVeg Tortonian run with ECHAM4/ML   107
Appendix B - List of symbols   121

Abstract

In this study, the climate of the Tortonian (Late Miocene, 11 to 7 Ma) and particularly the effects of the palaeovegetation on the climate are investigated using the complex atmospheric general circulation model ECHAM4 coupled to a mixed-layer ocean model (ML). Previous Tortonian simulations consider an adjusted palaeocean heat transport and an adapted palaeorography, but use the Recent vegetation (STEPPUHN, 2002; STEPPUHN ET AL., 2006; STEPPUHN ET AL., 2007). For the present Tortonian simulation, the palaeovegetation is considered in addition to the previously adapted Tortonian boundary conditions. A proxy-based reconstruction of the Tortonian vegetation is used to adapt the surface parameters in the ECHAM4/ML model and a Tortonian climate simulation is performed. According to this Tortonian run, the palaeovegetation has significant effects on the Late Miocene climate. Due to the adapted Tortonian vegetation, the meridional temperature gradient is reduced as compared to nowadays. The comparison with proxy data demonstrates, that an appropriate palaeovegetation contributes to a more realistic representation of the Tortonian climate in the model ECHAM4/ML. With model results of the Tortonian run with ECHAM4/ML, the carbon cycle and vegetation model CARAIB is run. In its main patterns, the simulated Tortonian vegetation of the CARAIB model agrees with the proxy-based reconstruction of the palaeovegetation. CARAIB sensitivity experiments demonstrate that variations in the atmospheric CO_2 are rather more important for the vegetation than differences between the Tortonian and today's climate. However, simulations with both models, ECHAM4/ML and CARAIB, are not completely in accordance with proxy data. Therefore, it can be concluded, that the Late Miocene climate is still not completely understood.

1 Introduction

Since the Cretaceous, the climate changed successively from a greenhouse world to the glacial and interglacial states of the Holocene (fig. 1.1). On the one hand, the Cenozoic cooling during the last 65 million years is quite well known from proxy data such as isotope records (PEARSON ET AL., 2001). On the other hand, modelling studies focus repeatedly on the Cretaceous (OTTO-BLIESNER & UPCHURCH, 1997; UPCHURCH ET AL., 1998; UPCHURCH ET AL., 1999) as well as on the glacials and interglacials of the Quaternary (GANOPOLSKI ET AL., 1998a; GANOPOLSKI ET AL., 1998b; KUBATZKI ET AL., 2000; LORENZ ET AL., 1996; MONTOYA ET AL., 1998). The warmer and more humid climate of the Cretaceous is caused by the configuration of continents (BARRON & WASHINGTON, 1984), an increased poleward oceanic and atmospheric heat transport (DECONTO ET AL., 2000; HERMAN & SPICER, 1996) and a high concentration of atmospheric CO_2 (BERNER, 1994). In contrast to this, the colder and variable Quaternary climate is primarily affected by orbital cycles (BERGER, 1978) and low concentrations of atmospheric CO_2 (JOUZEL ET AL., 1993).

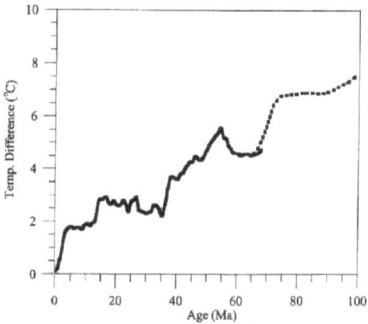

Figure 1.1: *The global average temperature from 100Ma to present with respect to today (modified from CROWLEY & ZACHOS, 2000). See CROWLEY & ZACHOS (2000) for details and original data sources.*

During the last couple of years, the successive Cenozoic cooling becomes also of interest to the community of climate modellers (DUTTON & BARRON, 1997; FLUTEAU ET AL., 1999; MIKOLAJEWICZ ET AL., 1993; STEPPUHN ET AL., 2006; STEPPUHN ET AL., 2007). The Miocene (23.8 to 5.3 Ma) is characterised as a transitional period from the Cretaceous greenhouse mode to the icehouse world of the Quaternary. Although the Miocene boundary conditions such as the land-sea distribution are basically comparable to nowadays, modelling studies (DUTTON & BARRON, 1997; STEPPUHN ET AL., 2006; STEPPUHN ET AL., 2007) and proxy data (BRUCH, 1998; WOLFE, 1994a) suggest a warmer and more humid climate than today. In particular, the meridional temperature gradient of the Miocene (fig. 1.2) is shallower than today (CROWLEY & ZACHOS, 2000; STEPPUHN ET AL., 2006).

Ocean modelling studies demonstrate, that the oceanic circulation during the Neogene differs significantly from present conditions (BICE ET AL., 2000; MAIER-REIMER ET AL., 1990; MIKOLAJEWICZ ET AL., 1993). Variations in the ocean circulation patterns are attributed to plate tectonic movements from the Neogene till today, which cause a different-than-today bathymetry as well as openings and closures of ocean gateways (BICE ET AL., 2000). Consequently the poleward heat transport in the oceans is affected (MIKOLAJEWICZ & CROWLEY, 1997). A weakening of the thermohaline circulation in the North Atlantic Ocean is caused by an open Central American Isthmus (MAIER-REIMER ET AL., 1990). This means that the northward heat transport in the North Atlantic Ocean is lower (MAIER-REIMER ET AL., 1990). During the Cenozoic, the closure of the Panama Isthmus leads to the development of the North Atlantic thermohaline circulation (MIKOLAJEWICZ & CROWLEY, 1997). In order to test the response of the atmosphere to a lower ocean heat transport, COVEY & THOMPSON (1989) apply an atmospheric general circulation model (AGCM) coupled to a mixed-layer ocean model. From this study, both atmospheric heat fluxes, the latent and the sensible heat flux, accomplish a higher northward heat transport, which partly compensates the weaker oceanic heat transport (COVEY & THOMPSON, 1989).

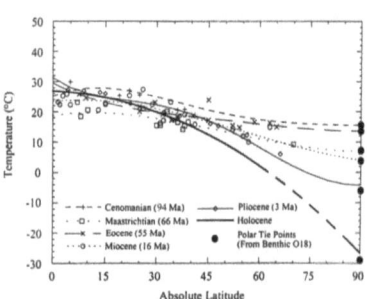

Figure 1.2: *The zonal average sea surface temperatures (modified from CROWLEY, 2000) for the present Holocene interglacial, Pliocene (3 Ma), Miocene (16 Ma), Eocene (55 Ma), Maastrichtian (66 Ma) and Cenomanian (94 Ma). See CROWLEY (2000) for details and original data sources.*

Concerning atmospheric modelling, Miocene studies investigate the climatic effects of a changed palaeogeography and palaeorography as compared to nowadays (BARRON, 1985; FLUTEAU ET AL., 1999; RAMSTEIN ET AL., 1997). From the Oligocene till today, the shrinking Paratethys contributes to a change from a warmer to a cooler climate in Asia (RAMSTEIN ET AL., 1997). The influence of the Paratethys is not only indicated for the formerly warm Siberian climate but also for Eastern Europe, which is more humid than today (RAMSTEIN ET AL., 1997).

The same model experiments (FLUTEAU ET AL., 1999; RAMSTEIN ET AL., 1997) include the uplift of Tibet during the Cenozoic. The reduced elevation of the Tibetan Plateau is responsible for an Asian monsoon, which is weaker than today (FLUTEAU ET AL., 1999; RAMSTEIN ET AL., 1997). However, these AGCM studies (FLUTEAU ET AL., 1999; RAMSTEIN ET AL., 1997) use Recent sea surface temperatures (SSTs).

Within a special research program (SFB 275 „Klimagekoppelte Prozesse in meso- und känozoischen Geoökosystemen") at the University of Tübingen (BERICHT DES SFB 275 DER UNIVERSITÄT TÜBINGEN), modelling studies with the AGCM ECHAM4 coupled to a mixed-layer ocean model (ML) focus on a realistic representation of the Tortonian (Late Miocene, 11 to 7 Ma). In order to understand the relevant climatic processes, the lower boundary conditions are successively adapted to the Tortonian (STEPPUHN, 2002). A first modelling approach considers a weaker palaeoceanic heat transport and a lower palaeorography (STEPPUHN ET AL., 2006). This Tortonian run reproduces some realistic climate tendencies, but also some discrepancies such as a too steep meridional temperature gradient (STEPPUHN ET AL., 2006). Because of these insufficiencies, STEPPUHN ET AL. (2007) perform a sensitivity study taking additionally a higher CO_2 concentration than previously used into account. Due to this CO_2 forcing, particularly the equator-to-pole temperature gradient is reduced, but the tropics are unrealistically warm (STEPPUHN ET AL., 2007). Thus, it is concluded that a higher-than-present CO_2 concentration cannot explain the differences between the Miocene and present-day's climate (STEPPUHN ET AL., 2007). For the Late Miocene, it is more likely to assume a CO_2 content which is as high as today or even lower (PEARSON & PALMER, 2000).

So far several studies focus on the effects of a changed palaeogeography and a lower palaeorography (FLUTEAU ET AL., 1999; RAMSTEIN ET AL., 1997) as well as on a weaker oceanic heat transport (BICE ET AL., 2000; MIKOLAJEWICZ ET AL., 1993; MIKOLAJEWICZ & CROWLEY, 1997; STEPPUHN ET AL., 2006). Nevertheless, the reasons particularly for the flat equator-to-pole temperature gradient of the Miocene are still poorly understood. For the above mentioned Tortonian studies within a PhD thesis (STEPPUHN, 2002), the modern vegetation is used (STEPPUHN ET AL., 2006; STEPPUHN ET AL., 2007). The proxy data base providing information about vegetation for the Miocene is rather scarce, but supports a different situation as today (MAI, 1995; WOLFE, 1994a; WOLFE, 1994b). During the Miocene, forests extend far towards the high latitudes (MAI, 1995; WOLFE, 1994a; WOLFE, 1994b). For the Early Miocene, a modelling study demonstrates, that a vegetation, which is characterised by a larger amount of forests, leads to a polar warming (DUTTON & BARRON, 1997). Therefore, the Early Miocene vegetation causes a reduced meridional temperature gradient (DUTTON & BARRON, 1997). However, DUTTON & BARRON (1997) use a rather simple vegetation reconstruction with only four biome types. If numerical models should contribute to an understanding of the Miocene climate, particularly focusing on the rather shallow meridional temperature gradient, detailed information about the Tortonian vegetation is needed.

In order to simulate a realistic Tortonian climate, the present PhD thesis continues the above mentioned Tortonian simulations with ECHAM4/ML (STEPPUHN ET AL., 2006, STEPPUHN ET AL., 2007) and considers an appropriate palaeovegetation. As the prior Tortonian runs (STEPPUHN ET AL., 2006; STEPPUHN ET AL., 2007) are used as the reference base for the new Tortonian run, which additionally includes an adjusted palaeovegetation, they are summarised in the following section. The proxy-based reconstruction of the Tortonian vegetation is presented in sec. 3, which is used for the new Tortonian run (sec. 4). Concerning the effects of the palaeovegetation, the analysis focuses on variations of the temperature and precipitation patterns and also on changes of the atmospheric circulation regimes. To test whether the adapted palaeovegetation contributes to a more realistic representation of the Tortonian climate, results of the new Tortonian run are quantitatively compared to terrestrial proxy data (sec. 5). The ECHAM4/ML output of the new Tortonian run is used for model simulations with the carbon cycle and vegetation model CARAIB (sec. 6). This allows to test the performance of the model simulations in terms of vegetation. Assuming different atmospheric CO_2 concentrations for the CARAIB runs, the relevance of variations in CO_2 on vegetation is estimated with respect to differences between the Tortonian and present-day's climate.

2 TORTONIAN REFERENCE SIMULATIONS WITH ECHAM4/ML

At the University of Tübingen, atmospheric modelling studies for the Late Miocene started within a special research program, the Sonderforschungsbereich (SFB) 275 „Klimagekoppelte Prozesse in meso- und känozoischen Geoökosystemen" (BERICHT DES SFB 275 DER UNIVERSITÄT TÜBINGEN). In a PhD thesis (STEPPUHN, 2002), the atmospheric general circulation model ECHAM4 coupled to a mixed-layer (ML) ocean model (DKRZ MODELLBETREUUNGSGRUPPE, 1994; DKRZ MODELLBETREUUNGSGRUPPE, 1997; ROECKNER ET AL., 1992; ROECKNER ET AL., 1996) is applied to the Tortonian for the first time. A major topic of these Tortonian simulations (STEPPUHN ET AL., 2006; STEPPUHN ET AL., 2007) is the realistic representation of the Late Miocene climate and the understanding of the relevant processes. For instance, the question, which processes contribute to a shallower-than-today Miocene meridional temperature gradient, is not sufficiently answered (CROWLEY & ZACHOS, 2000).

The present PhD thesis concentrates on the effects of the Tortonian vegetation. Concerning setup parameters (e.g. the paleoceanic heat transport), this investigation is however based on STEPPUHN's (2002) previous Tortonian model simulations, which are referred to as the *Standard Tortonian run* (STEPPUHN ET AL., 2006) and the $2 \times CO_2$ *Tortonian run* (STEPPUHN ET AL., 2007) in the following. These prior Tortonian runs are also used as reference runs for the new Tortonian simulation. In the following subsections, the Tortonian reference simulations are reinterpreted and summarised.

2.1 The model ECHAM4/ML

The atmospheric general circulation model ECHAM (DKRZ MODELLBETREUUNGSGRUPPE, 1994; DKRZ MODELLBETREUUNGSGRUPPE, 1997; ROECKNER ET AL., 1992; ROECKNER ET AL., 1996) is based on a weather prediction model (SIMMONS ET AL., 1989) of the European Centre for Medium Range Weather Forecast (ECMWF). For the purpose of global climate modelling, this model was advanced at the Max-Planck Institute (MPI) for Meteorology (ROECKNER ET AL., 1996). It includes original routines of the ECMWF model, but some parameterisations underwent modifications or were replaced (ROECKNER ET AL., 1996). ECHAM4 is a spectral model, which is based on the primitive equations. The prognostic variables are represented by a series of spherical harmonics. In the present Tortonian simulations, the series are truncated at wave number 30 (T30), which corresponds to a horizontal resolution of 3.75° × 3.75°. For the vertical, a hybrid sigma-pressure coordinate system with 19 layers is used. The model physics of ECHAM4 includes schemes for land surface processes, radiation, clouds, convection et cetera (DKRZ MODELLBETREUUNGSGRUPPE, 1994; DKRZ MODELLBETREUUNGSGRUPPE, 1997; ROECKNER ET AL., 1992; ROECKNER ET AL., 1996).

For the Tortonian simulations, the ECHAM4 model is coupled to a simple 50m-mixed-layer (ML) ocean model (STEPPUHN ET AL., 2006; ROECKNER, pers. comm.). This allows to describe the ocean heat transport without calculating the full ocean circulation. In order to perform simulations for the Tortonian, the ocean heat transport has to be adjusted. For this purpose, STEPPUHN ET AL. (2006) establish a new method, which is briefly explained below.

2.2 The Standard Tortonian run

2.2.1 The model setup of the Standard Tortonian run

In a first ECHAM4/ML simulation for the Tortonian, STEPPUHN ET AL. (2006) consider the effects of a generally weaker palaeoceanic heat transport and a generally lower Tortonian orography. For the setup of this Standard Tortonian run, the land-sea distribution of the Tortonian remains unchanged as compared to the modern one. This is justifiable, as only minor plate tectonic movements occur since the Late Miocene (PRELL & KUTZBACH, 1992; RAMSTEIN ET AL., 1997; RUDDIMAN & KUTZBACH, 1989). Furthermore, STEPPUHN ET AL. (2006) use the ECHAM model in its standard horizontal resolution of T30 (3.75°). This resolution is too coarse to consider small variations in the land-sea distriubution. However, the orography is adapted to the Tortonian. According to a global reconstruction of the palaeorography (KUHLEMANN, pers. comm.), the height of mountain ranges is generally reduced for the Standard Tortonian run. For example, Greenland reaches only about a tenth of its Recent elevation. The global vegetation represents modern conditions, except that the Recent Greenland ice cap is replaced by a tundra vegetation.

In order to adapt the heat transport of the mixed-layer ocean model to the Tortonian, a new method is presented by STEPPUHN ET AL. (2006). This method is based on the assumption, that the near-surface ocean circulation during the Tortonian is basically comparable to the modern pattern and that the meridional gradient of the sea surface temperatures (SSTs) is a measure for the oceanic heat transport. In order to obtain the palaeoceanic heat transport, local palaeo-SSTs are obtained from $\delta^{18}O$ data of foraminifera. From satellite observation data, a transfer function is determined to convert Recent local SSTs into Recent zonal average SSTs. This transformation is used to calculate the zonal average palaeo-SSTs from the local palaeo-SSTs. Assuming that the ratio of the meridional gradients of the Recent and palaeo-SSTs is a zonally constant value, the palaeo-flux correction of the mixed-layer ocean model is calculated. With this new method of STEPPUHN ET AL. (2006), it is possible to prescribe a global heat transport for a mixed-layer ocean model on the basis of a few data points. The reconstruction of palaeo-SSTs can be affected by several effects such as the diagenesis, but STEPPUHN ET AL.'s (2006) method is quite robust concerning small variations in the oxygen isotope composition.

The reconstructed Tortonian ocean heat transport is characterised by a generally weaker northward heat and mass transport (STEPPUHN ET AL., 2006). This is in accordance to studies with ocean general circulation model, which investigate the effects of an open Panama Isthmus during the Miocene (MIKOLAJEWICZ ET AL., 1993; MIKOLAJEWICZ & CROWLEY; 1997). During the Miocene, the open Panama Strait causes an exchange of the higher saline Atlantic Ocean water and the lower saline Pacific Ocean, which weakens the thermohaline circulation in the North Atlantic Ocean (MIKOLAJEWICZ & CROWLEY; 1997). Studies with OGCMs (BARRON & PETERSON, 1991; BICE ET AL., 2000; MAIER-RAIMER ET AL., 1990) and proxy data (COLLINS ET AL., 1996; FLOWER & KENNETT, 1994; HAUG & TIEDEMANN, 1998; TSUCHI, 1997; WOODRUFF & SAVIN, 1989) indicate, that changes in the oceanic circulation during the Neogene are caused by the changing ocean bathymetry, and by openings and closures of seaways such as the Central American Isthmus.

In addition to the above described boundary conditions, the atmospheric CO_2 in the Standard Tortonian run is specified with the present-day's level of 353 ppm. This lies within the spectrum of values which are given for the Miocene (CERLING ET AL., 1997, PAGANI ET AL., 1999; PEARSON & PALMER, 2000).

2.2.2 Results of the Standard Tortonian run

The global average temperature of the Standard Tortonian run (tab. 2.1) is almost unaffected (–0.1 °C) as compared to the one of the *Recent Control run* (available from the DKRZ Modellbetreuungsgruppe). The difference in the global average precipitation between the Standard Tortonian and the Control run is rather small (–4 mm/a). In the Standard Tortonian run, a decreased sea ice volume is shown (tab. 2.1). The sea ice volume is more reduced on the Northern than on the Southern Hemisphere in the Standard Tortonian run as compared to the Recent Control experiment.

From the zonal average temperature (fig. 2.1), the northern high latitudes are shown to be warmer (+2 °C) in the Standard Tortonian run than in the Control run. In the lower latitudes, the zonal average temperatures of both runs, the Standard Tortonian and the Control run, do not differ much. Thus, the equator-to-pole temperature difference is reduced in the Standard Tortonian run. From the horizontal temperature pattern (fig. 2.2a), the highest increase rates as compared to the Control run are observed for Greenland (+15 °C) and the Himalayan (+20 °C). These warming effects are primarily due to the lowering of the orography and the removal of continental ice sheets in the Standard Tortonian run. In the case of the lower Himalayan, the reduced annual precipitation rate (–400 mm/a) indicates a weakened Asian monsoon in the Standard Tortonian run (fig. 2.2b).

In contrast to the warmer-than-present high latitudes, the Standard Tortonian run indicates that mean annual temperatures decrease in the mid-latitudes (–1 °C to –2 °C), which is

	T_s [°C]	p_{tot} [mm/a]	Sea Ice Volume [×10^{12} m³]			
			Northern Hemisphere		Southern Hemisphere	
			JJA	DJF	JJA	DJF
Recent Control run	15.7	1020	11.3	17.8	17.5	14.8
Standard Tortonian run	15.6	1016	9.4	16.3	16.6	13.8
2×CO_2 Tortonian run	18.6	1051	1.5	6.1	12.6	10.2

Table 2.1: *The global average surface temperature T_s [°C], the global average precipitation p_{tot} [mm/a], and the Northern and Southern Hemisphere's seasonal average sea ice volume [×10^{12} m³] of the Recent Control run, the Standard Tortonian run (STEPPUHN ET AL., 2006) and the 2×CO_2 Tortonian run (STEPPUHN ET AL., 2007).*

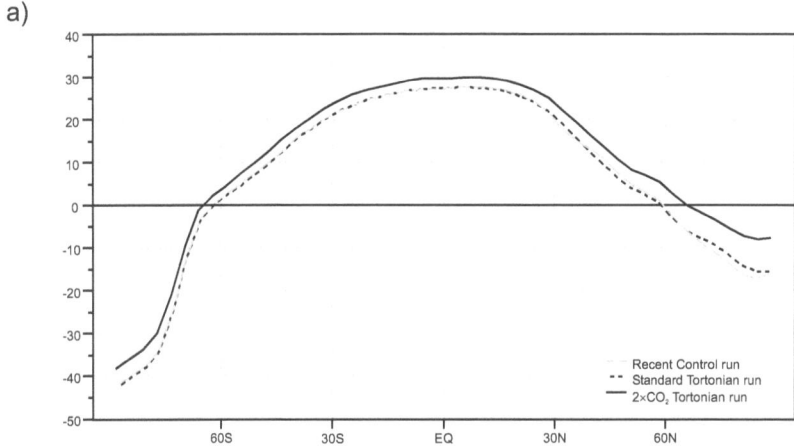

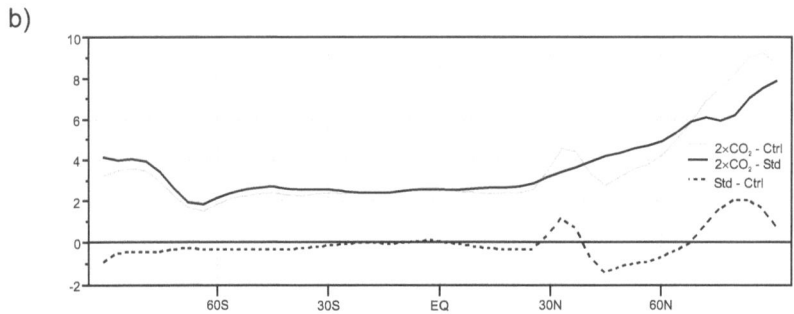

Figure 2.1: a) The zonal average temperature [°C] of the Recent Control run (**grey dashed**), the Standard Tortonian run (**black dotted**) and the 2×CO_2 Tortonian run (**black solid**). b) The zonal average temperature difference [°C] between the 2×CO_2 Tortonian and the Control run (**grey solid**), the 2×CO_2 Tortonian and the Standard Tortonian run (**black solid**) and the Standard Tortonian run and the Control run (**black dashed**).

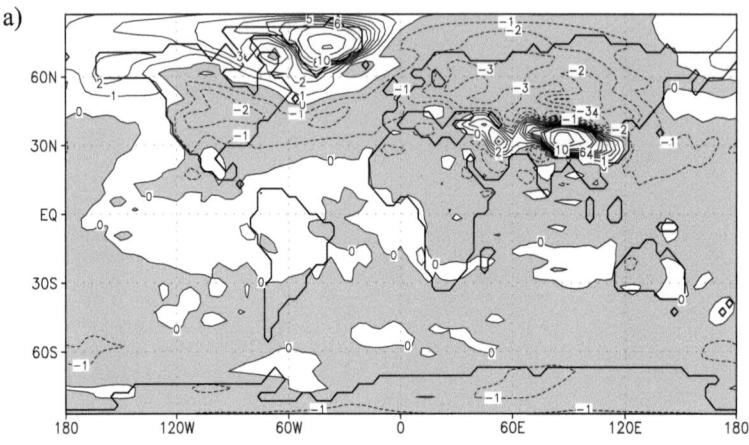

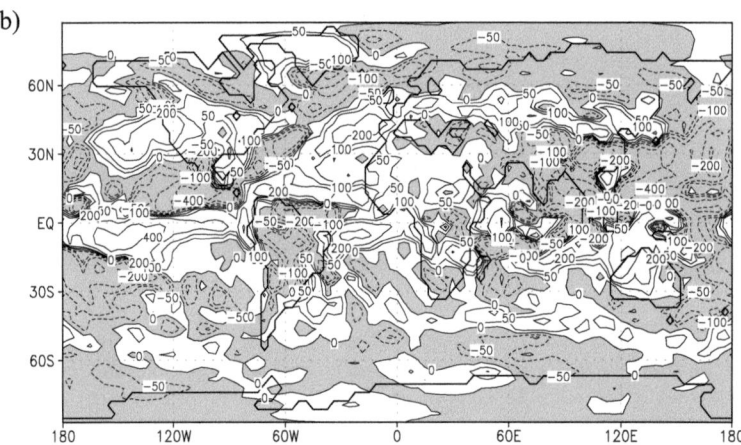

Figure 2.2: *The difference between the Standard Tortonian run and the Recent Control run for a) the mean annual 2m-temperature [°C] and b) the annual precipitation [mm/a] (modified from STEPPUHN ET AL., 2006). Negative values are shown grey shaded.*

attributed to the weaker palaeoceanic heat transport. In Southern Europe, higher precipitation rates as compared to the Control experiment (fig. 2.2b) indicate more frequent storm activities in the Standard Tortonian run. In the low latitudes, the atmospheric circulation in the Standard Tortonian run differs from the one in the Recent Control experiment (STEPPUHN, 2002). In the Pacific region, the temperature anomaly pattern (fig. 2.2a) between the Standard Tortonian run and the Recent Control simulation is comparable to an El Niño (STEPPUHN ET AL., 2006).

2.2.3 Verification of the Standard Tortonian run

The Standard Tortonian run basically agrees with other modelling studies as well as with proxy data (STEPPUHN ET AL., 2006). The reduced meridional temperature gradient in the Standard Tortonian run agrees with an Early Miocene simulation of DUTTON & BARRON (1997) using the AGCM GENESIS. However, the simulation of DUTTON & BARRON (1997) demonstrates warmer conditions and a more flattened equator-to-pole temperature gradient than the Standard Tortonian run. As compared to nowadays, the amount of sea ice is slightly lower in the Standard Tortonian run (tab. 2.1), but still more ice cover as indicated by proxy data (WOLF & THIEDE, 1991). Terrestrial proxy data (BRUCH, 1998; WOLFE, 1994b) suggest warmer conditions in the high and mid-latitudes than observed from the Standard Tortonian run. For the lower latitudes, the Asian monsoon is weaker in the Standard Tortonian run than today. Other modelling studies (FLUTEAU ET AL., 1999; RAMSTEIN ET AL., 1997) and proxy data (WU ET AL., 1998) also mention a weaker-than-today Miocene monsoon. Concerning the Pacific region, an El Niño-like pattern of the Standard Tortonian run with respect to the Control run can be supposed. In the eastern Pacific region, the Standard Tortonian run demonstrates surface temperature anomalies of +1 °C to +2 °C (STEPPUHN ET AL., 2006). From modern remote sensing, temperature anomalies of +2 °C to +4 °C in the Pacific region are observed during an El Niño (WEBSTER & PALMER, 1997). Since using a mixed-layer ocean model, the El Niño phenomenon itself cannot be modelled (STEPPUHN ET AL., 2006).

Except for some discrepancies, the Standard Tortonian run reflects that the large-scale trends qualitatively agree quite well with proxy data (summarised in tab. 2.2). Some minor discrepancies in the Standard Tortonian run can be ascribed to the Paratethys, which is not taken into account (STEPPUHN ET AL., 2006). AGCM experiments demonstrate, that the shrinking Paratethys contributes to the Cenozoic cooling in Central Eurasia (RAMSTEIN ET AL., 1997). However, the patterns of the Standard Tortonian run are basically consistent with other modelling studies which include a Paratethys (RAMSTEIN ET AL., 1997). It is concluded that either an underestimated atmospheric CO_2 or the vegetation is responsible for the major discrepancies in the Standard Tortonian run. For the Miocene, various estimations of the atmospheric CO_2 concentration exist (PAGANI ET AL., 1999; PEARSON & PALMER, 2000; VAN DER BURGH ET AL., 1993). For the CO_2 level during the Miocene, an upper limit of 700 ppm is suggested from CERLING (1991).

2.3 The $2 \times CO_2$ Tortonian run

2.3.1 The model setup of the $2 \times CO_2$ Tortonian run

In order to test the hypothesis whether a possibly underestimated CO_2 is responsible for the above mentioned insufficiencies of the Standard Tortonian run, a sensitivity experiment is performed (STEPPUHN ET AL., 2007). For this $2 \times CO_2$ *Tortonian run*, the same boundary conditions as for the Standard Tortonian run are used, but the CO_2 concentration is doubled (= 700ppm).

2.3.2 Results of the $2 \times CO_2$ Tortonian run

The $2 \times CO_2$ Tortonian run demonstrates a global warming (+3 °C) and an increased global average precipitation (+35 mm/a) as compared to the Standard Tortonian run (tab. 2.1). Moreover, the double CO_2 simulation represents much warmer and more humid conditions as the Control run (tab. 2.1). Because of the globally warmer conditions, the sea ice is reduced (tab. 2.1) in the $2 \times CO_2$ Tortonian run as compared to the Standard Tortonian run and the Recent Control run. The loss of Arctic sea ice is relatively more during northern summer (–87 %) than during northern winter (–66 %) in the $2 \times CO_2$ Tortonian run as compared to today. During summer, the Arctic sea ice almost vanishes in the $2 \times CO_2$ Tortonian run. The absolute reduction of the Arctic sea ice volume is higher during the winter months than during summer. This indicates a decreased seasonal contrast due to the doubled CO_2 concentration.

Focusing on the zonal average temperatures (fig. 2.1), an increase across all latitudes is observed in the $2 \times CO_2$ Tortonian run as compared to the Standard Tortonian run. The strongest warming occurs in the high northern latitudes (+8 °C) in the $2 \times CO_2$ Tortonian run. This contributes to the reduction of the Arctic sea ice volume (tab. 2.1) as well as to a flattened meridional temperature gradient in the $2 \times CO_2$ Tortonian run.

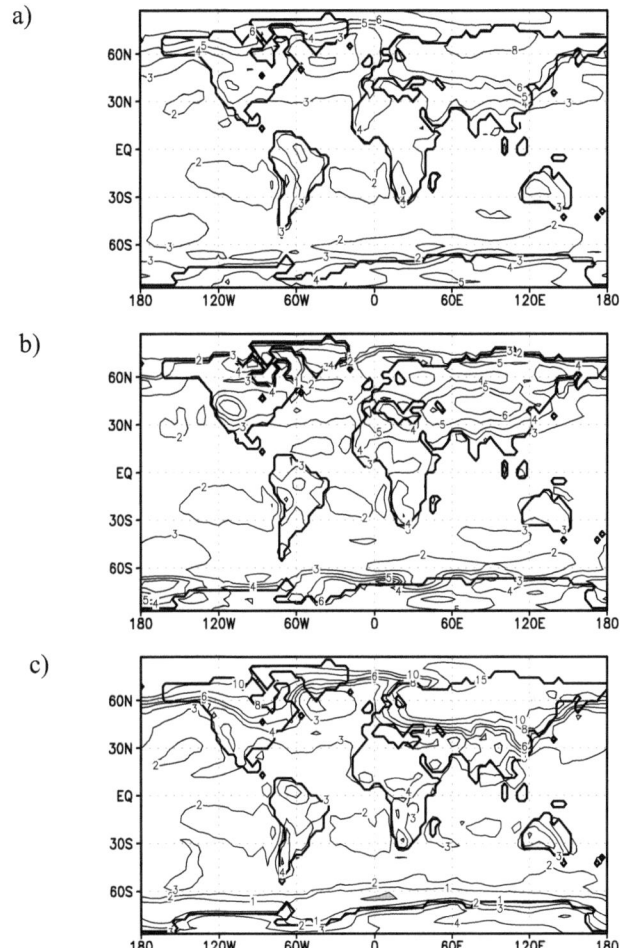

Figure 2.3: *The average 2m-temperature anomalies [°C] between the 2×CO_2 Tortonian run and the Standard Tortonian run for a) the annual average, b) JJA and c) DJF (modified from* STEPPUHN ET AL., *2007). Negative values are shown grey shaded.*

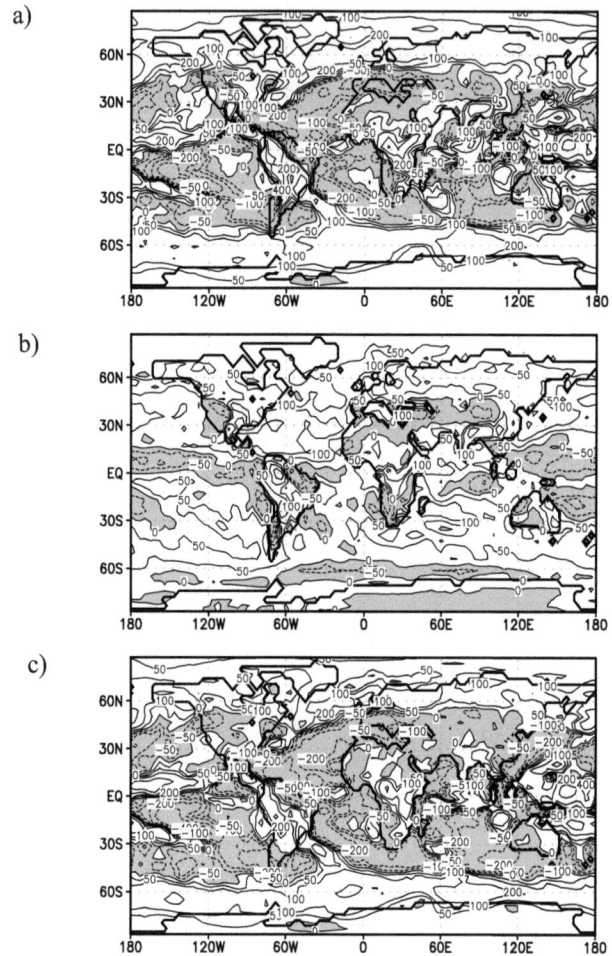

Figure 2.4: *The anomalies between the 2×CO_2 Tortonian and the Standard Tortonian run for a) the annual precipitation [mm/a], b) the annual evapotranspiration [mm/a] and c) the difference between the annual precipitation and evapotranspiration [mm/a] (modified from STEPPUHN ET AL., 2007). Negative values are shown grey shaded.*

Due to the different heat capacity, the temperature increase in the 2×CO_2 Tortonian run with respect to the Standard Tortonian run (fig. 2.3a) is more pronounced over continents than over ocean surfaces (+2 °C). Over the continents, the doubled CO_2 causes warmer conditions of more than +5 °C in the annual average. In particular, the high latitudes of the Northern Hemisphere are much warmer (+8 °C) in the 2×CO_2 scenario as compared to the Standard Tortonian run. Focusing on the seasonal temperature differences between both Tortonian runs (fig. 2.3b,c), the CO_2-induced warming is stronger during winter than during summer and the Northern Hemisphere is more affected than the Southern. During northern summer in the northern mid-latitudes (fig. 2.3b), the 2×CO_2 Tortonian run demonstrates the highest warming rates of about +7 °C. In the high latitudes of the Northern Hemisphere, the summerly temperature difference between the 2×CO_2 Tortonian and the Standard Tortonian run is about +3 °C to +5 °C. During the winter season of the Northern Hemisphere (fig. 2.3c), this pattern is reversed. From the low to the mid-latitudes, winter temperatures raise modestly by about +3 °C in the 2×CO_2 Tortonian run. In the high latitudes, the 2×CO_2 Tortonian run shows winterly temperature increases of more than +15 °C as compared to the Standard Tortonian run. This pattern is a result of a strong ice-albedo feedback in the high latitudes in the 2×CO_2 Tortonian run (cf. tab2.1). This feedback reduces the seasonal temperature cycle as compared to the Standard Tortonian run.

With a globally increased temperature in the double CO_2 scenario, the hydrological cycle is also intensified (cf. tab. 2.1), as a warmer atmosphere stores more water vapour. Over continental areas, the annual precipitation increases in the 2×CO_2 Tortonian run (fig. 2.4a). Focusing on Asia, an increase in precipitation of +200 mm/a is observed from the Tortonian CO_2 run. Therefore, the monsoon is enhanced, which is attributed to the doubling of CO_2. In the tropics, the highest absolute increase rates (+400 mm/a) occur in the 2×CO_2 Tortonian run as compared to the Standard Tortonian run. In the high latitudes, the precipitation increase is also relatively high (+100 mm/a) in the 2×CO_2 Tortonian run, as the total amount of precipitation is generally lower than in lower latitudes.

The globally warmer conditions in the 2×CO_2 Tortonian run lead to an increased evapotranspiration rate (fig. 2.4b). Over continents, the evapotranspiration is generally higher in the 2×CO_2 Tortonian run as compared to the Standard Tortonian run (+100 mm/a). Considering anomalies of precipitation minus evapotranspiration (fig. 2.4c), the ocean surfaces of the tropical and subtropical latitudes are more arid (–400 mm/a) in the 2×CO_2 Tortonian run. In contrast to this, the oceans of higher latitudes are more humid (+200 mm/a) than in the Standard Tortonian run. In the 2×CO_2 Tortonian run, this results in a strengthened atmospheric moisture transport from the low towards the high latitudes over the oceans and from the oceans towards the land surfaces as compared to the Standard Tortonian run.

2.3.3 The comparison of model results and quantitative terrestrial proxy data

In order to validate whether the Standard Tortonian run or the 2×CO_2 Tortonian run is more realistic, results of both model runs are quantitatively compared with terrestrial proxy data (cf. sec. 5 for further details regarding the method and the data base). The comparison of the mean annual temperatures (MAT) demonstrates, that the Standard Tortonian run (fig. 2.5a) represents cooler conditions in the high latitudes than suggested by proxy data (WOLFE, 1994b). Contrarily, the MATs of the 2×CO_2 Tortonian run globally agree quite well to proxy-based estimations (fig. 2.6a). For Siberia, the double CO_2 scenario indicates lower temperatures than suggested from proxy data. For the mid-latitudes, the Standard Tortonian run tends to be cooler than the proxy estimations (fig. 2.5a), whereas the 2×CO_2 Tortonian run demonstrates a quite good agreement with proxy data (fig. 2.6a). In the lower latitudes, the 2×CO_2 Tortonian agrees worse with proxy data than the Standard Tortonian run. In Asia, the 2×CO_2 Tortonian run indicates a discrepancy to proxy data of more than +5 °C. For the Standard Tortonian run, the difference to the proxy-based estimation is about +3 °C in Asia. In the 2×CO_2 Tortonian run, the tropics and subtropics are therefore represented too warm as compared to proxy data.

From the annual precipitation (fig. 2.5b and 2.6b), both Tortonian runs globally agree quite well with the proxy-based estimations. For tropical regions, both model runs demonstrate much too humid conditions (more than +2000 mm/a), which indicate deficits in simulating realistic precipitation rates by the ECHAM model. With regard to Europe, both Tortonian runs agree with proxy data. In the 2×CO_2 Tortonian run, Southern Europe is too dry (–350 mm/a to –590 mm/a). North America is more arid (–400 mm/a) in the 2×CO_2 Tortonian run than terrestrial proxy data suggest. Regarding the annual precipitation, the Standard Tortonian run is more realistic than the 2×CO_2 Tortonian run.

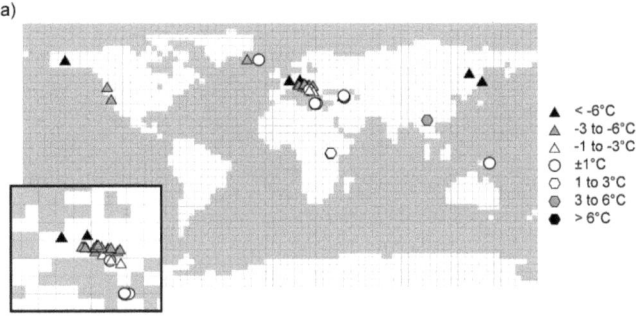

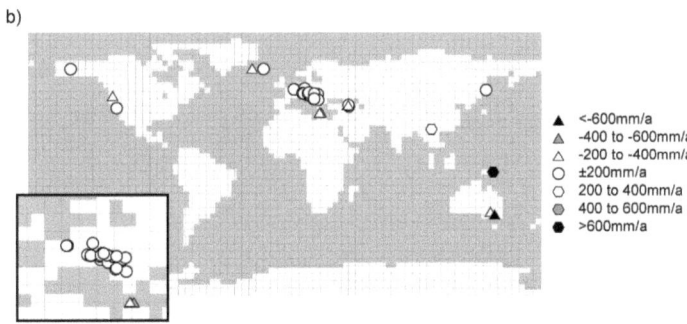

Figure 2.5: *The differences between the Standard Tortonian run and terrestrial proxy data for a) the mean annual temperature [°C], and b) the annual precipitation [mm/a] (modified from* STEPPUHN ET AL., *2007). The European region is shown enlarged. White circles represent consistency, triangles represent cooler or more arid conditions, and sexangles represent warmer or more humid conditions in the model simulation, respectively.*

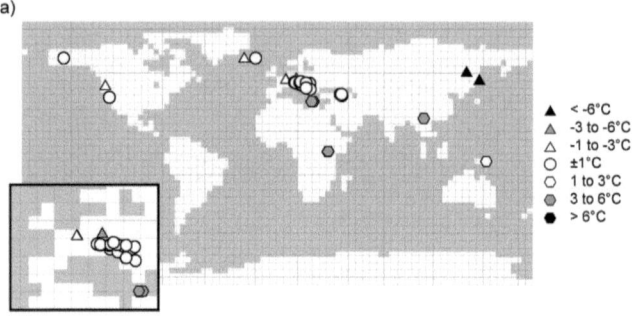

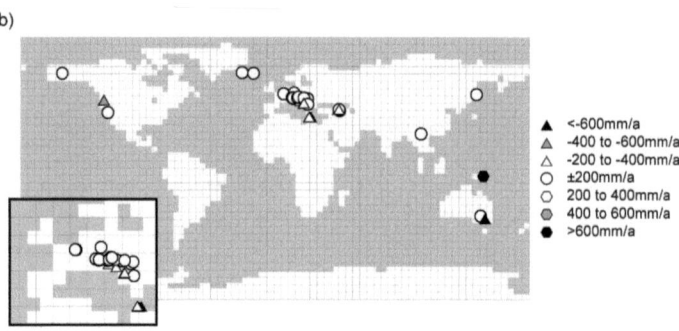

Figure 2.6: *The differences between the 2×CO$_2$ Tortonian run and terrestrial proxy data for a) the mean annual temperature [°C], and b) the annual precipitation [mm/a] (modified from STEPPUHN ET AL., 2007). The European region is shown enlarged. White circles represent consistency, triangles represent cooler or more arid conditions, and sexangles represent warmer or more humid conditions in the model simulation, respectively.*

2.3.4 Verification of the 2×CO$_2$ Tortonian run

Basically, the 2×CO$_2$ Tortonian run as compared to the Standard Tortonian run demonstrates the same global warming (+3.0 °C) as observed for Recent climate scenarios (+2.5 °C to +4 °C) by using different AGCMs coupled to mixed-layer ocean models (McGuffie et al., 1999). This indicates, that the feedback to the CO$_2$ forcing is almost independent from the different prescribed boundary conditions of the Recent climate studies and the 2×CO$_2$ Tortonian run.

The seasonal cycle in the 2×CO$_2$ Tortonian run is reduced as compared to the Standard Tortonian run. For the Holocene, a less distinct seasonal cycle, correlated with a lower amount of sea ice, is also shown (Berger et al., 1998). A sensitivity study with a prescribed lower Arctic sea ice amount supports a reduced contrast between summer and winter temperatures on the Northern Hemisphere (Raymo et al., 1990).

Focusing on the water cycle, the 2×CO$_2$ Tortonian run indicates more arid conditions over the subtropical Atlantic Ocean. Contrarily, the conditions over the North Atlantic Ocean are more humid. A higher CO$_2$ should therefore affect the North Atlantic thermohaline circulation. Recent climate studies, which use ocean general circulation models under high CO$_2$ conditions (Mikolajewicz & Voss, 2000; Rahmstorf, 1995), report a decreased surface salinity due to an increased freshwater input in the North Atlantic Ocean. This dilution weakens the North Atlantic Deep Water formation (Mikolajewicz & Voss, 2000; Rahmstorf, 1995). At the same time, the circulation in the Atlantic Ocean is strengthened due to an increased evaporation in the subtropics (Mikolajewicz & Voss, 2000). In total, the CO$_2$-induced freshwater changes in Recent model experiments, which are comparable to the patterns in the 2×CO$_2$ Tortonian run, tend to weaken the thermohaline circulation in the Atlantic Ocean (Mikolajewicz & Voss, 2000). However, since using a mixed-layer ocean model for the Tortonian simulations, an effect on the ocean circulation cannot be demonstrated.

On the regional scale, an enhanced Asian monsoon precipitation and a warming of +3 °C to +4 °C over South Asia is observed in the 2×CO$_2$ Tortonian run as compared to the Standard Tortonian run. A weakening of the Asian monsoon is demonstrated from the Standard Tortonian run when compared to the Recent Control run (Steppuhn et al., 2006) as well as from other modelling studies (Fluteau et al., 1999; Ramstein et al., 1997) and proxy data (Wu et al., 1998). Thus, the monsoon patterns are unrealistic in the 2×CO$_2$ Tortonian run than in the

Standard Tortonian run. However, modern climate models reveal uncertainties in representing the Asian monsoon (DOUVILLE ET AL., 2000).

The comparison of both Tortonian runs with terrestrial proxy data demonstrates a good agreement of the mean annual temperatures between the 2×CO_2 Tortonian run and proxy data (summarised in tab. 2.2). The tropics and subtropics are exceedingly too warm in the 2×CO_2 Tortonian as compared to proxy-based estimations. Consequently, a higher-than-present pCO_2 solves some problems of the Standard Tortonian run, but new difficulties arise. STEPPUHN ET AL. (2007) conclude, that the main insufficiencies in the Standard Tortonian run cannot be explained by an underestimated pCO_2.

For the 2×CO_2 Tortonian run, the atmospheric CO_2 concentration is 700 ppm, which is the doubled present-day's concentration of 353 ppm as used for the Standard Tortonian run. For the Miocene, various estimations of the atmospheric CO_2 content exist (CERLING ET AL., 1997; PAGANI ET AL., 1999; PEARSON & PALMER, 2000). On the one hand, the pCO_2 of the Standard Tortonian run lies within the range of CO_2 values. On the other hand, the 2×CO_2 Tortonian run represents the upper limit. For the Miocene, CERLING ET AL. (1997) suggest an atmospheric CO_2 concentration of more than 500ppm. From CERLING (1991), a Miocene pCO_2 between present-day's level and 700 ppm is estimated. A Late Miocene value of 380 to 400 ppm is estimated from VAN DER BURGH ET AL. (1993). PAGANI ET AL. (1999) suggest a CO_2 content equal to the

	Standard Tortonian run (STEPPUHN ET AL., 2006)	2×CO_2 Tortonian run (STEPPUHN ET AL., 2007)
global temperature	–	+
- high latitudes	– –	O
- mid-latitudes	– –	O
- low latitudes	O	+ +
precipitation	–	+
Arctic sea ice volume	+ +	O
meridional temperature gradient	+ +	O

Table 2.2: *The summarised qualitative agreements (O) and the over-/underestimations (+/–) of the model results of the Standard Tortonian run (STEPPUHN ET AL., 2006) and the 2×CO_2 Tortonian run (STEPPUHN ET AL., 2007) as compared to proxy data.*

pre-industrial level (280 ppm). A Miocene atmospheric CO_2 concentration, which is as high as the Recent level or even lower, is favoured by PEARSON & PALMER (2000).

The shallow meridional temperature gradient of the Miocene is one of the major problems of the Standard Tortonian run (cf. tab. 2.2). A higher-than-present CO_2 level leads to a reduction of the pole-to-equator gradient, which results in an unrealistically representation of the tropics (cf. tab. 2.2). Typical greenhouse climates such as the Cretaceous are (amongst others) controlled by a high CO_2 level (BERNER, 1994). However, climatic differences between the Miocene and today basically cannot be ascribed to variations of the atmospheric pCO_2. Thus, the causes for the differences between the modern and the Miocene climate are still not identified.

Both model simulations, the Standard Tortonian and the $2 \times CO_2$ Tortonian run, are based on the modern vegetation. However, Miocene proxy data indicate a different situation (MAI, 1995; WOLFE, 1994a; WOLFE, 1994b). During the Early Miocene, boreal forests extend far towards high latitudes and less deserts exist (DUTTON & BARRON, 1997; WOLFE, 1985). Sensitivity studies (DUTTON & BARRON, 1996) demonstrate a significant influence of vegetation on the climate system. For the Early Miocene, the palaeovegetation contributes to a flattening of the temperature gradient (DUTTON & BARRON, 1997). For the Cretaceous, this effect is also observed from sensitivity experiments (OTTO-BLIESNER & UPCHURCH, 1997; UPCHURCH ET AL., 1998). Thus, an appropriate palaeovegetation has to be further tested concerning its implications for a more realistic representation of the Tortonian climate.

3 The reconstruction of the Tortonian vegetation

3.1 The method

In order to perform a realistic Tortonian simulation with ECHAM4/ML (cf. sec. 4.1), detailed information about the Tortonian vegetation is demanded for each grid point of the ECHAM model (horizontal resolution of 3.75°). Based on palaeobotanical data such as fossil pollen and leaf data, and fossil carpoflora, UHL & BRUCH (pers. comm.) reconstruct the global Tortonian vegetation. The information about the Tortonian floras is provided from studies of BRUCH (1998), GRAHAM (1998), GREGOR (1982), GREGOR & UNGER (1988), JACOBS (1999), JACOBS & DEINO (1996), MAI (1995), MARTIN (1990), MARTIN (1998), MAY ET AL. (1999), MUDIE & HELGASON (1983), PLAYFORD (1982), SACHSE (unpublished data), SACHSE & MOHR (1996), UNGER (1983), WOLFE (1994a), WOLFE (1994b). The proxy data are interpreted and classified into biomes (PRENTICE ET AL., 1992; tab. 3.2). Since the attribution of fossil floras to a special biome type is not well-defined, the number of biomes is reduced. Some related biomes are grouped into a combined biome: cool & cold mixed forest, cool grass & tundra, and hot, cool & sand desert.

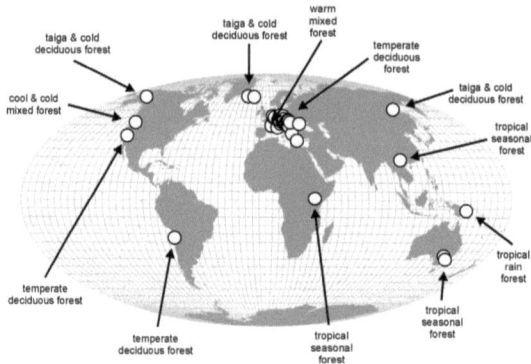

Figure 3.1: *The proxy data locations and the attributed Tortonian vegetation (see text for data sources).*

Scarce or even no proxy data are available for some continents such as Southern Africa or Latin America (fig. 3.1). Due to this lack of information, the Prentice biome model (PRENTICE AT AL., 1992) is applied by using climate data of the Standard Tortonian simulation (STEPPUHN ET AL., 2006 [cf. sec. 2.2]). From monthly average data (temperature, precipitation and cloud cover) of the Standard Tortonian run, the following parameters are derived: The minimum and maximum temperature of the coldest month ($T_{c,min}$ and $T_{c,max}$), the minimum temperature of the warmest month ($T_{w,min}$), the temperature sum of days above 0 °C and 5 °C (*gdd0* and *gdd5*) and a minimum and maximum moisture index (α_{min} and α_{max}), which is the ratio of the actual and the potential evapotranspiration (CLAUSSEN, pers. comm.; CLAUSSEN, 1993). According to tab. 3.1, the Prentice biome model claculates plant functional types (PFTs). Considering the dominance hierarchy (*D*), these PFTs define biomes (tab. 3.2). Finally, a global vegetation distribution is obtained from the Prentice biome model. As the previous Standard Tortonian run represents a slightly unrealistic climate (STEPPUHN ET AL., 2006; STEPPUHN ET AL., 2007), the calculated palaeovegetation differs from the proxy-based palaeovegetation. Based on the

	$T_{c,min}$	$T_{c,max}$	$T_{w,min}$	gdd5	gdd0	α_{min}	α_{max}	D
Trees								
Tropical evergreen	15.5					0.80		1
Tropical raingreen	15.5					0.45	0.95	1
Warm-temperate evergreen	5.0					0.65		2
Temperate summergreen	−15.0	15.5		1200		0.65		3
Cool-temperate conifer	−19.0	5.0		900		0.65		3
Boreal evergreen conifer	−35.0	−2.0		350		0.75		3
Boreal summergreen			5.0	350		0.65		3
Non-trees								
Xerophytic woods/shrub	5.0					0.28		4
Warm grass/shrub			22.0			0.18		5
Cool grass/shrub				500		0.33		6
Cold grass/shrub					100	0.33		6
Hot desert/shrub			22.0					7
Cool desert/shrub					100			8
Polar desert								9

Table 3.1: *The climatic restrictions for the plant functional types of the Prentice biome model (PRENTICE ET AL., 1992). $T_{c,min}$ and $T_{c,max}$ denote the minimum and maximum temperature of the coldest month [°C], $T_{w,min}$ the minimum temperature of the warmest month [°C], gdd5 and gdd0 denotes the temperature sum of days above 5 °C and 0 °C [°C], α_{min} and α_{max} is the minimum and maximum available moisture index and D denotes the dominance hierarchy.*

single locations where the palaeovegetation is known from proxy data, the calculated biome distribution is used to interpolate the palaeovegetation between these single locations. Fig. 3.2a shows the resulting reconstruction of the global Tortonian vegetation.

In order to quantify changes between the reconstructed Tortonian palaeovegetation and the modern distribution, the present-day's vegetation is also calculated with the Prentice biome model. From the Intergovernmental Panel on Climate Change (IPCC), monthly observation data (NEW ET AL., 1999) of the years 1961 to 1990 are taken to calculate the modern vegetation. This observation data set is available in the high horizontal resolution of 0.5°. The climatological 30-year-averages of temperature, precipitation and cloud cover are applied to the Prentice biome model. Fig. 3.2b shows the calculated modern vegetation distribution. Antarctica is not plotted, as the IPCC observation data set contains no information about this region.

Plant functional types	Biome
Tropical evergreen	Tropical rain forest
Tropical evergreen + tropical raingreen	Tropical seasonal forest
Tropical raingreen	Savanna
Warm-temperate evergreen	Warm mixed forest
Temperate summergreen + cool-temperate conifer + boreal summergreen	Temperate deciduous forest
Temperate summergreen + cool-temperate conifer + boreal evergreen conifer + boreal summergreen	Cool mixed forest
Cool-temperate conifer + boreal evergreen conifer + boreal summergreen	Cool conifer forest
Boreal evergreen conifer + boreal summergreen	Taiga
Cool-temperate conifer + boreal summergreen	Cold mixed forest
Boreal summergreen	Cold deciduous forest
Xerophytic woods/shrub	Xerophytic woods/shrub
Warm grass/shrub	Warm grass/shrub
Cool grass/shrub + cold grass/shrub	Cool grass/shrub
Cold grass/shrub	Tundra
Hot desert/shrub	Hot desert
Cool desert/shrub	Cool desert
Polar desert	Ice/polar desert

Table 3.2: *The allocation of plant functional types (cf. tab. 3.1) to biomes according to the Prentice biome model (PRENTICE ET AL., 1992).*

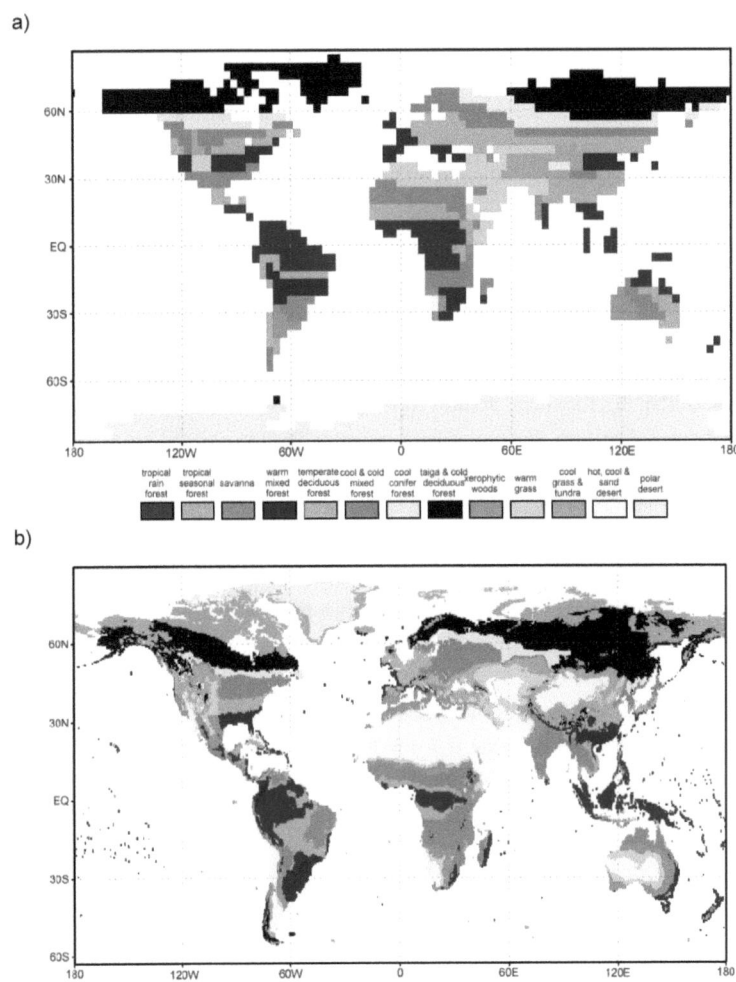

Figure 3.2: *a) The proxy-based reconstructed Tortonian vegetation, and b) the present-day's vegetation as calculated from IPCC observation data (NEW ET AL., 1999) using the biome model of PRENTICE ET AL. (1992).*

To obtain some information about the fractional portions, the percentage cover of the biomes is determined for the modern and the palaeovegetation. The Tortonian Antarctic continent is excluded from this comparison, as there is no information about Antarctica from the IPCC data set (NEW ET AL., 1999). Fig. 3.3a shows the fractional portions of the Recent and the Tortonian vegetation for each biome. In addition, the summed fractional portions of forest types (biomes no. 1 to 8), grassland types (no. 9 to 11) and deserts (no. 12 and 13) are represented in fig. 3.3b.

3.2 The resulting Tortonian vegetation

As shown from fig. 3.2, the Tortonian palaeovegetation is generally more dense than nowadays. During the Tortonian, tropical forests (no. 1 to 3) expand and their margins shift further poleward. The Late Miocene Africa is generally characterised by a tropical forest vegetation. Accordingly, the Sahara desert is smaller than today and consists of steppe and open grassland rather than sand desert. A more woody Tortonian vegetation replaces the present-day's warm-arid desert, semi-desert and grassland regions. It is assumed that formerly no extreme sand deserts comparable to the modern Sahara desert exist. During the Tortonian, the extension of tropical forests (no. 1 to 3) is +5 % larger than today (fig. 3.3). Deserts diminish by −17 % as compared to the present-day's situation.

For the mid-latitudes (fig. 3.2), the tendency towards denser forests (no. 4 to 8) in the Tortonian is also observed. From fig. 3.3, the fractional portions of forest types no. 4 to 8 globally increase by +1.5 % to +5 %. During the Tortonian, warm mixed forests and temperate deciduous forests replace the modern Central European forests. Cool and cold mixed forests, which represent the modern Central European vegetation, are found further northward as compared to today. During the Tortonian, cool conifer forests as well as taiga & cold deciduous forests occur far into the high latitudes. In particular, Greenland is assumed to be largely covered with taiga & cold deciduous forests instead of present-day's ice cover. During the Late Miocene, a tundra vegetation covers about 4 % of the global land surface (mainly northern parts of Asia and North America). This is a reduction of −7 % as compared to nowadays

During the Tortonian, the extension of all forest types except for savanna (no. 3) and taiga & cold deciduous forest (no. 8) is almost doubled in relation to the absolute present-day's values. The relative loss of the sparse vegetation cover is in the same order of magnitude. This represents not only a simple shift of climate zones: Since the Tortonian (or even earlier), deserts and semi-deserts grow at the expense of forest vegetation. For the modern warm-arid and cold-continental areas, the Tortonian surface properties such as the albedo and the leaf area index have to differ clearly from today's conditions.

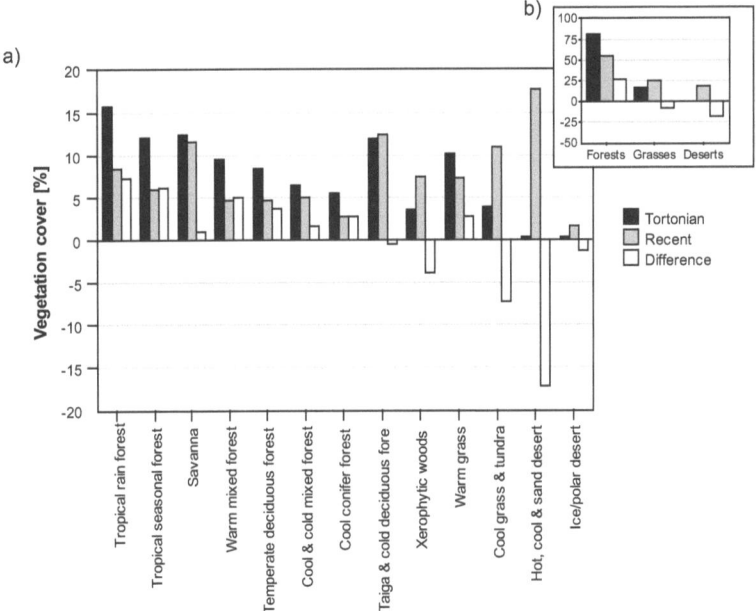

Figure 3.3: *The fractional cover [%] of the Tortonian vegetation (black), the Recent vegetation (grey) and their differences (white) a) for each biome, and b) for the summed forest, grassland and desert types.*

4 THE PALVEG TORTONIAN RUN WITH ECHAM4/ML

4.1 The model setup

As previously emphasised (cf. sec. 2), this study does not only concentrate on a more realistic simulation of the Tortonian climate, but also on the climatic implications of an adapted Tortonian palaeovegetation. It continues the previous Tortonian runs (cf. sec. 2, STEPPUHN ET AL., 2006; STEPPUHN ET AL., 2007). The model ECHAM4/ML (DKRZ MODELLBETREUUNGSGRUPPE, 1994; DKRZ MODELLBETREUUNGSGRUPPE, 1997; ROECKNER ET AL., 1992; ROECKNER ET AL., 1996) is used in its standard resolution of T30 (3.75°). The same Tortonian boundary conditions as described for the Standard Tortonian run (cf. sec. 2.2.1, STEPPUHN ET AL., 2006) are used, but not the modern vegetation. According to the reconstruction of the Tortonian vegetation (cf. sec. 3), land surface parameters are adapted. To consider the changed vegetation in the ECHAM model, data for the albedo (α_v) the leaf area index (LAI) the vegetation and forest cover (c_{veg} and c_{for}, respectively) and the maximum soil water capacity ($W_{s,max}$) are required (tab. 4.1). For each biome, these values are taken from CLAUSSEN (1994), except for the maximum available soil water capacity which is taken from HAGEMANN ET AL. (1999). For the re-grouped biomes such as cool grass & tundra, the average of both biome values is used. For the Tortonian, it is assumed, that extreme sand deserts do not occur (cf. fig. 3.2a). To complete the listing in tab. 4.1, the surface parameters of the biome type sand desert are also specified.

4.2 Model results of the PalVeg Tortonian run as compared to the Standard Tortonian run

For the Tortonian, an ECHAM4/ML model run is performed using the above described boundary conditions. In the following, this simulation is referred to as the *PalVeg Tortonian run*. After 10 of 20 simulation years, the model reaches an equilibrium state and the last 10 years are taken into account for further analysis. In order to compare the results of the PalVeg Tortonian run, data of the Standard Tortonian run (STEPPUHN ET AL., 2006) and the *Recent Control run* of ECHAM4/ML (experiment EXP700-run712, available from the DKRZ Modellbetreuungsgruppe) are considered. Data of each run are averaged over the last 10

#	Biome	α_v	LAI	c_{veg}	c_{for}	$W_{s,max}$
1	Tropical rain forest	0.12	9.3	0.98	0.98	0.360
2	Tropical seasonal forest	0.12	4.3	0.82	0.82	0.200
3	Savanna	0.15	2.6	0.65	0.58	0.695
4	Warm mixed forest	0.15	6.0	0.91	0.79	0.300
5	Temperate deciduous forest	0.16	2.7	0.65	0.65	0.233
6	Cool & cold mixed forest	0.15	2.0	0.54	0.54	0.140
7	Cool conifer forest	0.13	9.1	0.97	0.97	0.380
8	Taiga & cold deciduous forest	0.14	3.7	0.77	0.77	0.161
9	Xerophytic woods	0.18	2.6	0.66	0.19	0.480
10	Warm grass	0.20	0.8	0.27	0	0.680
11	Cool grass & tundra	0.18	1.1	0.35	0.03	0.213
12	Hot & cool desert	0.20	0.3	0	0	0.100
	(Sand desert	0.35	0	0	0	0.100)
13	Ice/polar desert	0.15	0	0	0	0.035

Table 4.1: *The allocation of surface parameters as used for the PalVeg Tortonian run with ECHAM4/ML. α_v designates the albedo [frac.], LAI the leaf area index [m^2/m^2], c_{veg} and c_{for} the vegetation and forest cover [frac.], respectively, and $W_{s,max}$ the maximum available soil water capacity [m]. Values are modified from* CLAUSSEN *(1994) and* HAGEMANN ET AL. *(1999).*

simulation years. Appendix A includes the relevant absolute data fields of the PalVeg Tortonian run. To figure out the effects of vegetation on the Late Miocene climate, the following figures show the differences between both Tortonian model runs, the PalVeg Tortonian minus the Standard Tortonian run. In order to decide if the climate of the PalVeg Tortonian run is significantly distinct from the Standard Tortonian run, a 'Student t-test' (VON STORCH & ZWIERS, 1999) is performed. Variables at each grid point are assumed to be nearly normal distributed (*Central Limit Theorem*). The statistical test determines whether the mean difference of both runs is significantly distinct from zero (*null hypothesis*). The level of significance is set to $p = 0.05$. For the figures shown in the following, non-significant differences between the model runs are (dark) grey shaded (e.g., fig. 4.2). In some selected figures (e.g., fig. 4.5), the light grey shading indicates negative values, which is useful to easily identify specific patterns.

At first, the data description of the results concentrates on the global average differences of temperature, precipitation and sea ice of the PalVeg Tortonian run with respect to the Standard Tortonian and the Recent Control run (sec. 4.2.1). The following subsections focus on the zonal average temperatures (sec. 4.2.2) and the regional anomaly patterns (sec. 4.2.3) between the PalVeg and the Standard Tortonian simulation. The description of vegetation-induced climatic changes is completed with the analysis of large-scale (sec. 4.2.4) and regional-scale (sec. 4.2.5) variations in the atmospheric circulation between the PalVeg Tortonian run and the Standard Tortonian run. Finally, the PalVeg Tortonian run is compared to the modern Control run regarding the regional temperature pattern (sec. 4.3.1) and the water cycle. The water cycle includes the global (sec. 4.3.2) and zonal averages (sec. 4.3.3) as well as the regional (sec. 4.3.4) precipitation and evapotranspiration patterns.

4.2.1 The global average temperature, precipitation and sea ice

Tab. 4.2 represents the global averages of the surface parameters as preset for the Standard Tortonian run and the Recent Control run, and the PalVeg Tortonian run. During the Tortonian, particularly forests occur, which extend far into the high latitudes (cf. sec. 3.2). Accordingly, the vegetation and forest cover increase (+66 % and +124 %, respectively) as compared to nowadays. The albedo decreases by –40 % in the PalVeg Tortonian run. Comparing the values of tab. 4.2 with tab. 4.1, the Recent Control run and the Standard Tortonian run represent globally sparse vegetation. For instance, the global land albedo ($\alpha_v = 0.20$) corresponds to desert and warm grass conditions. Contrarily, the PalVeg Tortonian run demonstrates a global land albedo of tropical or coniferous forests ($\alpha_v = 0.12$). Thus, globally increased temperatures are expected in the PalVeg Tortonian run. The maximum available soil water capacity (+8 %)

	α_v	LAI	c_{veg}	c_{for}	$W_{s,max}$
Standard Tortonian run & Recent Control run	0.20	1.8	0.32	0.21	0.247
PalVeg Tortonian run	0.12	3.2	0.53	0.48	0.267
Difference	–0.08	+1.5	+0.21	+0.26	+0.020

Table 4.2: *The global averages (land surface grid points only are considered) of the albedo α_v [frac.], the leaf area index LAI [m²/m²], the vegetation and forest cover c_{veg} and c_{for} [frac.], and the maximum available soil water capacity $W_{s,max}$ [m] of the Standard Tortonian run (and the Recent Control run), the PalVeg Tortonian run and the difference between the model simulations.*

and the leaf area index (+83 %) increase in the PalVeg Tortonian run. Hence, it is expected that the climatic conditions are more humid in the PalVeg Tortonian run as compared to the Standard Tortonian run and the Recent Control run.

From tab. 4.3, the PalVeg Tortonian run demonstrates indeed a globally increased temperature as compared to the Recent Control run (+0.6 °C) and the Standard Tortonian run (+0.9 °C). The temperature raise is higher on the Northern Hemisphere (+1.2 °C) than on the Southern (+0.6 °C) as compared to the Standard Tortonian run. For the seasonal average temperatures of the Northern Hemisphere, a stronger increase in the PalVeg Tortonian run is apparent during winter. The Northern Hemisphere's seasonality (the difference between summer and winter temperatures) is not reduced significantly (–0.1 °C) in the PalVeg Tortonian run as compared to the Recent Control run. Contrarily, a vegetation-induced reduction of the seasonality is observed (–0.6 °C) in relation to the Standard Tortonian run. On the Southern Hemisphere, the summer and the winter temperatures increase (+0.6 °C) if the palaeovegetation is considered. Correspondingly, the Southern Hemisphere's conditions are warmer in the PalVeg Tortonian run, but the seasonality is not noticeably affected as compared to the Standard Tortonian run.

	T_{2m} [°C]							P_{tot} [mm/a]
	global	Northern Hemisphere			Southern Hemisphere			global
		annual	JJA	DJF	annual	JJA	DJF	
Recent Control run	15.3	15.5	20.9	9.9	15.1	13.5	16.9	1019
Standard Tortonian run	15.0	14.9	21.0	9.5	15.0	13.1	16.8	1010
PalVeg Tortonian run	15.9	16.1	21.9	11.0	15.6	13.7	17.4	1046
Standard - Recent	–0.3	–0.6	+0.1	–0.4	–0.1	–0.4	–0.1	–9
PalVeg - Recent	+0.6	+0.6	+1.0	+1.1	+0.5	+0.2	+0.5	+27
PalVeg - Standard	+0.9	+1.2	+0.9	+1.5	+0.6	+0.6	+0.6	+36

Table 4.3: *The average near-surface temperature, T_{2m} [°C], the global annual average, the annual and seasonal average temperature for both hemispheres, respectively, and the global annual precipitation, p_{tot} [mm/a] of the Recent Control run, the Standard Tortonian run, the PalVeg Tortonian run and the differences between the model simulations, respectively.*

Regarding the annual precipitation (tab. 4.3), the PalVeg Tortonian run demonstrates an enforced hydrological cycle. The global precipitation increases by +27 mm/a in the PalVeg Tortonian run as compared to the Recent climate scenario. The Standard Tortonian run represents a slightly drier situation than today (–9 mm/a). Thus, the palaeovegetation causes an increase in the global average precipitation (+36 mm/a).

According to the warmer conditions of the PalVeg Tortonian run, the northern sea ice volume gets smaller (tab. 4.4). During summer as well as during winter, the Arctic sea ice volume is reduced (-4×10^{12} m^3) as compared to the Control run. As compared to the Standard Tortonian run, the Arctic ice volume is diminished by -2.6×10^{12} m^3 during winter and by -2.1×10^{12} m^3 during summer in the PalVeg Tortonian simulation. The reduction of the Arctic sea ice is relatively more intense during summer (–22 %) than during winter (–16 %) in the PalVeg Tortonian run as compared to the Standard Tortonian run.

4.2.2 The zonal average temperature

The zonally averaged mean annual temperatures (fig. 4.1) indicate constantly warmer conditions in the PalVeg Tortonian run than in the Standard Tortonian run. Again, the Northern Hemisphere is more affected than the Southern. The annual increase in temperature on the

	Sea Ice Volume [×10^{12} m^3]			
	Northern Hemisphere		Southern Hemisphere	
	Summer (JJA)	Winter (DJF)	Summer (DJF)	Winter (JJA)
Recent Control run	11.3	17.8	14.8	17.5
Standard Tortonian run	9.4	16.2	13.9	16.7
PalVeg Tortonian run	7.3	13.7	14.6	17.0
Standard - Recent	–1.9	–1.6	–0.9	–0.8
PalVeg - Recent	–4.0	–4.1	–0.2	–0.5
PalVeg – Standard	–2.1	–2.5	+0.7	+0.3

Table 4.4: *The seasonal average sea ice volume for the Northern and Southern Hemisphere [×10^{12} m^3] of the Recent Control run, the Standard Tortonian run, the PalVeg Tortonian run and the differences between the model simulations, respectively.*

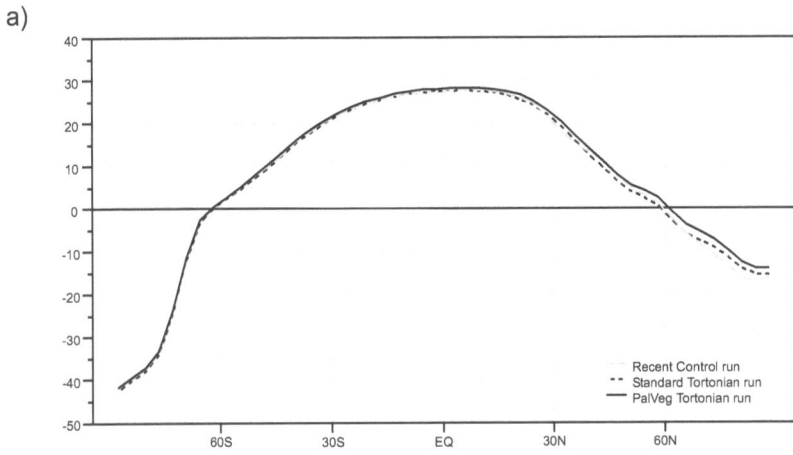

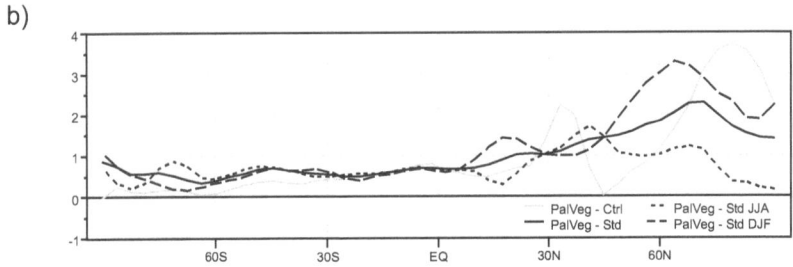

Figure 4.1: a) The zonal average temperatures [°C] of the Recent Control run (**grey dashed**), the Standard Tortonian run (**black dotted**) and the PalVeg Tortonian run (**black solid**). b) The difference of the zonal average temperatures [°C] of the annual average of the PalVeg Tortonian minus the Control run (**grey solid**), the annual average of the PalVeg Tortonian minus the Standard Tortonian run (**black solid**), the average of the PalVeg Tortonian minus the Standard Tortonian run for northern summer (**black dotted**) and the average of the PalVeg Tortonian minus the Standard Tortonian run for northern winter (**black dashed**).

Southern Hemisphere is about +0.5 to +0.7 °C and not varying with latitudes (cf. tab. 4.3) in the PalVeg Tortonian run with respect to the Standard Tortonian run. On the Northern Hemisphere, the temperature differences between both Tortonian runs vary with latitude. Starting at the equator and proceeding further towards 70°N, the temperature differences increase steadily between the PalVeg Tortonian and the Standard Tortonian run. At around 70°N, a maximum difference of +2.3 °C is reached. Further poleward, the temperature differences between both Tortonian runs decrease but remain positive. Thus, the warming in the PalVeg Tortonian run with respect to the Standard Tortonian run is stronger in higher latitudes than in mid- and low latitudes. The meridional temperature gradient is reduced in the PalVeg Tortonian run. In this context it is important to note that the palaeocean heat transport is adjusted with palaeo-SSTs, which are obtained from $\delta^{18}O$ data (cf. sec. 2.2.1; STEPPUHN ET AL., 2006). These palaeo-SSTs represent a shallower equator-to-pole gradient. Thus, the pattern of the PalVeg Tortonian run is consistent with the pattern of the reconstructed palaeo-SSTs. The reduction of the polar sea ice in the PalVeg Tortonian run (cf. sec. 4.2.1) is a further consequence of the vegetation-induced warmer high latitudes.

Considering the seasonal pattern of the PalVeg Tortonian run as compared to the Standard Tortonian run (fig. 4.1b), the summerly and winterly warming of the Southern Hemisphere is almost the same as for the annual average (+0.6 °C). On the Southern Hemisphere, seasonal contrasts are not reduced in the PalVeg Tortonian run. In the low and in the high latitudes of the Northern Hemisphere, the PalVeg Tortonian run demonstrates larger temperature increases during winter than during summer as compared to the Standard Tortonian run. In the low and in the high latitudes, the seasonality is reduced in the PalVeg Tortonian run. In the lower mid-latitudes at around 40°N, the seasonality contrarily increases in the PalVeg Tortonian run as summer temperatures raise more than winter temperatures.

4.2.3 The regional temperature, precipitation and evapotranspiration patterns

Temperature

Comparing the PalVeg with the Standard Tortonian run, the horizontal pattern of the mean annual temperature (fig. 4.2a) indicates larger increases in temperature over land surfaces than over the oceans. Over the oceans, the temperatures rise moderately (+0.5 °C) in the PalVeg Tortonian run. For North Africa, an increase in temperature of +3 °C is observed from the

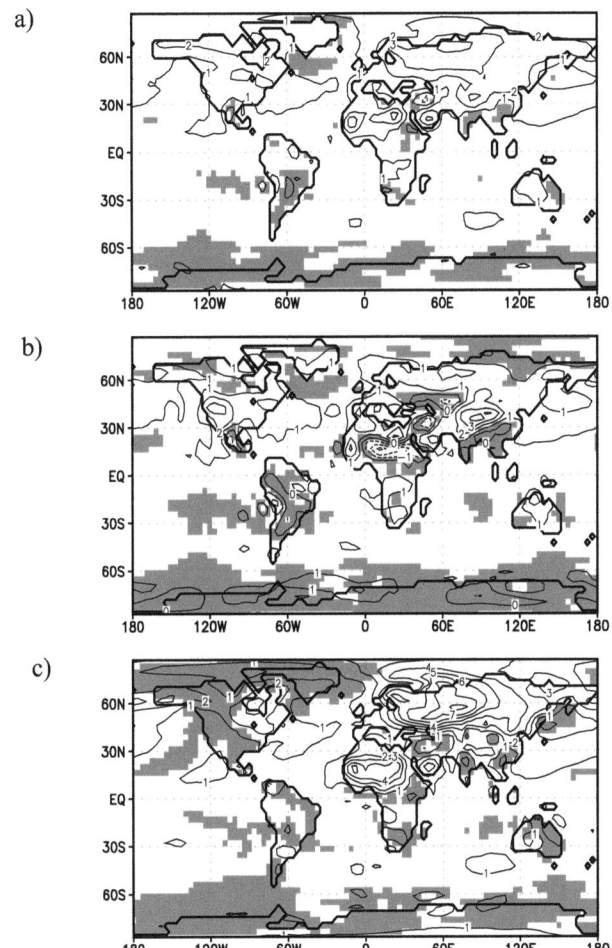

Figure 4.2: *The average near-surface 2m-temperature anomalies [°C] between the PalVeg Tortonian and the Standard Tortonian run for a) the annual average, b) JJA and c) DJF. Shaded areas represent non-significant anomalies with a local Student's t-test (p = 0.05).*

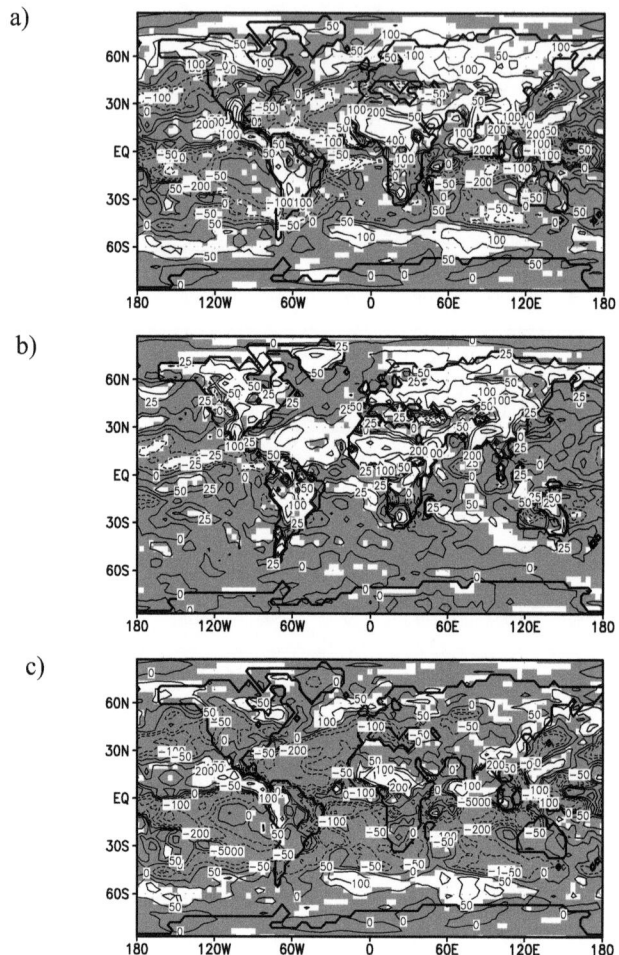

Figure 4.3: *The annual average anomalies between the PalVeg Tortonian and the Standard Tortonian run of a) the total precipitation [mm/a], b) the evapotranspiration [mm/a] and c) the total precipitation minus the evapotranspiration [mm/a]. Shaded areas represent non-significant anomalies with a local Student's t-test (p = 0.05).*

PalVeg Tortonian run. For North America and Eurasia/Asia, a raise of +4 °C is demonstrated as compared to the Standard Tortonian run. On the Northern Hemisphere, temperatures increase more than on the Southern as compared to the Standard Tortonian run. This is a result of the different landmass distribution between both hemispheres.

Regarding the seasonal average temperatures (fig. 4.2b,c) of the Southern Hemisphere, the summerly (DJF) and winterly (JJA) situation is about the same as described for the annual average pattern of the PalVeg Tortonian run as compared to the Standard Tortonian run. On the Northern Hemisphere, significant seasonal temperature anomalies between both Toprtonian runs are demonstrated, which differ from the annual average pattern. For the area of the Recent Sahara desert, the PalVeg Tortonian run indicates a summerly cooling of more than –2 °C. In North Africa during northern winter (DJF), a temperature increase of about +4 °C occurs as compared to the Standard Tortonian run. Thus, the seasonal contrast in North Africa is smaller in the PalVeg Tortonian run. For the northern parts of Eurasia, winter temperatures rise by maximally +7 °C in the PalVeg Tortonian run, whereas it is just about +1 °C during summer. For the high latitudes of Eurasia, a significant reduction of the seasonality is demonstrated in the PalVeg Tortonian run.

Precipitation and evapotranspiration

The annual total precipitation rate is shown in fig. 4.3a. As mentioned above (cf. sec. 4.2.1), the changed vegetation induces the tendency to a more humid atmosphere in the PalVeg Tortonian run. This is similar to an enforced hydrological cycle as compared to the Standard Tortonian run. In the tropics, the rainfall over continents intensifies significantly in the PalVeg Tortonian run. Particularly in Southeast Asia and in the recently warm-arid area of the Sahara desert the rainfall raises more than +400 mm/a as compared to the Standard Tortonian run. For the cold-continental region of Siberia, the precipitation is +100 mm/a higher in the PalVeg Tortonian run. For the high latitudes, the precipitation increase in the PalVeg Tortonian run is relatively high as the absolute rates are lower than in the tropics.

For the Tortonian, the vegetation is generally more dense than today (cf. sec. 3). Hence, it is expected that not only the precipitation increases in the PalVeg Tortonian run (cf. sec. 4.2.1) but also the evapotranspiration (fig. 4.3b). Over the continents, large increases in evapotranspiration are demonstrated from the PalVeg Tortonian run (fig. 4.3b). North Africa and Southeast Asia indicate the highest increase rates (+400 mm/a) as compared to the Standard Tortonian run. In the mid- and high latitudes, the intensified rainfall (+100 mm/a) in the PalVeg Tortonian run is also noticeable. There, the relative amplification with respect to the Standard Tortonian run is remarkable as the absolute evapotranspiration is lower than in tropical regions (cf. precipitation).

Over the tropical Atlantic Ocean, the precipitation minus the evapotranspiration, $p_{tot} - E$, (fig. 4.3c) indicates a water deficit in the PalVeg Tortonian run as compared to the Standard Tortonian runs. Over the Northern Atlantic Ocean, the precipitation increases noticeably more than the evapotranspiration in the PalVeg Tortonian run. This demonstrates an increased moisture transport from the tropics towards higher latitudes over the Atlantic Ocean in the PalVeg Tortonian run. Regarding the land surfaces, the resulting difference of precipitation and evapotranspiration is also of interest. For Southeast Asia, the PalVeg Tortonian simulation demonstrates more humid conditions (+100 mm/a). For large areas, the precipitation (fig. 4.3a) as well as the evapotranspiration (fig. 4.3b) rates are higher in the PalVeg Tortonian run (cf. above), but $p_{tot} - E$ (fig. 4.3c) is almost zero between both Tortonian runs. Consequently, there is an intensified internal turnover of moisture in the PalVeg simulation.

4.2.4 The large-scale atmospheric circulation patterns

For all atmospheric layers, the atmosphere is generally warmer (fig. 4.4a) and more humid (fig. 4.4b) in the PalVeg Tortonian run. Particularly at around 20°N, the specific humidity demonstrates a near-surface increase of more than +1 g/kg, which is attributed to the palaeovegetation. Corresponding to the temperature (fig. 4.4a) and humidity (fig. 4.4b) pattern, the total heat flux from the surface to the atmosphere (fig. 4.4c,d) increases in the PalVeg Tortonian run. However, changes in the sensible heat flux are of minor importance (fig. 4.4c) in the PalVeg Tortonian run. The more dense palaeovegetation releases primarily an additional amount of latent heat energy into the atmosphere. Considering the radiation budget (fig. 4.4e,f), the changes in the sensible and latent heat flux in the PalVeg Tortonian run correlate with a generally positive radiation budget as compared to the Standard Tortonian run. For some latitudes (e.g., at around 10°N), the solar radiation flux (fig. 4.4e) is lower in the PalVeg Tortonian run as compared to the Standard Tortonian run. The increase in the atmospheric water content (fig. 4.4b) causes higher precipitation rates (fig. 4.3a), which are equal to an increased cloud cover and thus a higher absorption of solar radiation (fig. 4.4e) in

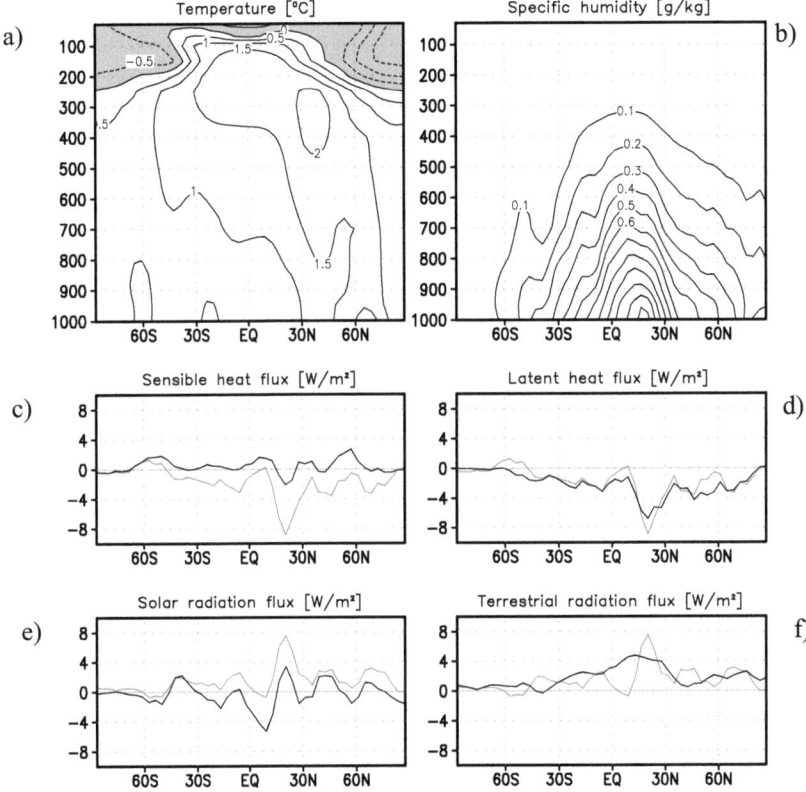

Figure 4.4: *The zonal and annual average differences between the PalVeg Tortonian run and the Standard Tortonian run for a) the temperature [°C] with respect to the height (in pressure coordinates [hPa]), b) the specific humidity [g/kg] with respect to the height (in pressure coordinates [hPa]), c) the surface sensible heat flux [W/m^2]* (**black**), *d) the surface latent heat flux [W/m^2]* (**black**), *e) the surface solar radiation flux [W/m^2]* (**black**) *and f) the surface terrestrial radiation flux [W/m^2]* (**black**). *The grey curves in c) and d) represent the total (sensible + latent) surface heat flux differences [W/m^2] and in e) and f) the total (solar + terrestrial) surface radiation flux differences [W/m^2] between the PalVeg Tortonian run and the Standard Tortonian run. The contour intervals are 0.5 °C for a) and 0.1 g/kg for b). Negative values of the temperature are grey shaded in a).*

the PalVeg Tortonian run. On the one hand, the lower albedo (e.g., in North Africa at around 20°N) partly compensates the reduced incoming solar radiation flux in the PalVeg Tortonian run (fig. 4.4e). On the other hand, the solar radiation deficit is compensated by a decreased terrestrial radiation outflux (fig. 4.4f) as compared to the Standard Tortonian run. For almost all latitudes, more heat is transported from the surface into the atmosphere in the PalVeg Tortonian run.

A higher atmospheric energy turnover in the PalVeg Tortonian run affects the atmospheric circulation patterns as compared to the Standard Tortonian run (fig. 4.5). The global circulation in the PalVeg Tortonian run is intensified as compared to the Standard Tortonian run (fig. 4.5). From the zonally averaged zonal wind component (fig. 4.5a), a generally intensified zonal mass flow is indicated in the PalVeg Tortonian run. Regarding the vertical circulation cells such as the Hadley cell, the PalVeg Tortonian simulation demonstrates a poleward shift of these cells (fig. 4.5c; cf. also fig. A.6 in the Appendix A). The surface pressure patterns (fig. 4.6) support the poleward shift of the vertical circulation cells in the PalVeg Tortonian run. As indicated from fig. 4.4c,d, this shift is caused by the additional energy release from the surface to the atmosphere, which produces an upward component of the vertical wind in the PalVeg Tortonian run (fig. 4.5c). It should be noted that because of the pressure coordinate system negative anomalies of the vertical wind component (fig. 4.5c) denote upward movements.

The zonal average vertical wind component (fig. 4.5c) demonstrates that particularly the down branch of the Northern Hemisphere's Hadley cell shifts northward for slightly less than one grid cell as compared to the Standard Tortonian run. Due to the resolution of the ECHAM model (3.75°), a more exact specification of this northward shift is not reasonable. The northern Hadely cell enlarges for more than 300km in its meridional extension in the PalVeg Tortonian run. The Ferrel and the Polar cell are slightly smaller (fig. 4.5c) as compared to the Standard Tortonian run. The meridional wind (fig. 4.5b) indicates a weakening of the Hadley circulation in the PalVeg Tortonian run. Within the Ferrel and the Polar cell, the meridional mass and heat transport increases due to the palaeovegetation (fig. 4.5b). Particularly in the mid- and high latitudes, the intensified meridional transport in the PalVeg Tortonian run contributes to a reduction of the equator-to-pole temperature gradient (cf. sec. 4.2.2) as compared to the Standard Tortonian run.

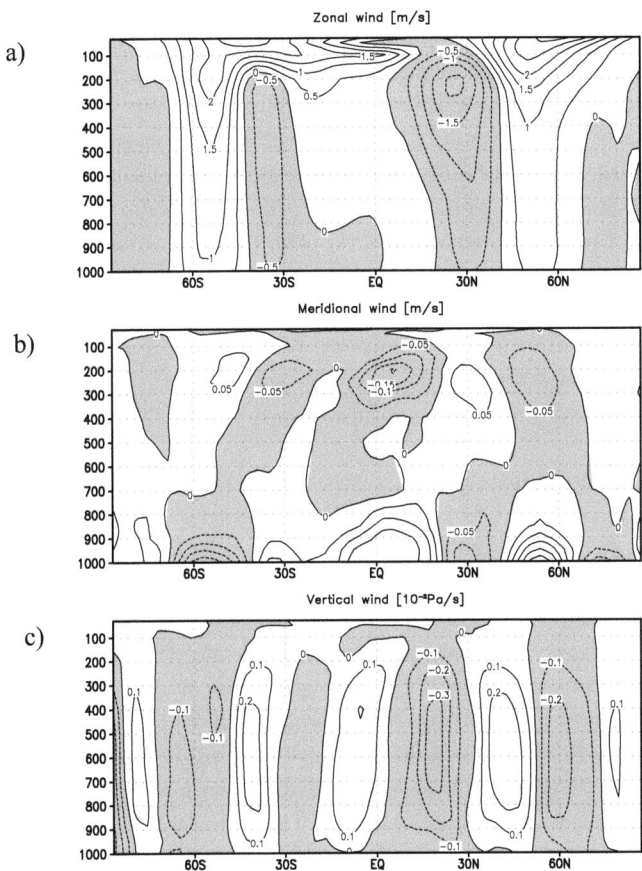

Figure 4.5: *The zonal average differences of a) the zonal wind [m/s], b) the meridional wind [m/s] and c) the vertical wind [10^{-2} Pa/s] between the PalVeg Tortonian run and the Standard Tortonian run with respect to the height (in pressure coordinates [hPa]). The contour intervals 0.5 m/s for a), 0.05 m/s for b), and 0.1×10^{-2} Pa/s for c), respectively. Negative values are grey shaded.*

4. THE PALVEG TORTONIAN RUN WITH ECHAM4/ML

As previously mentioned, the surface pressure (fig. 4.6) supports the poleward shift of the vertical circulation cells in the PalVeg Tortonian run with respect to the Standard Tortonian run. In addition, the surface pressure indicates a more zonal circulation in the PalVeg Tortonian run. The near-surface (1000 hPa) wind field (fig. 4.7a) also demonstrates the intensified zonal mass flow of the PalVeg Tortonian run. On both hemispheres, the westward trade winds of the lower latitudes and the westerlies of the mid-latitudes are more intense than in the Standard Tortonian run. Considering higher atmospheric levels, this pattern is also backed by the wind speed field (fig. 4.8) and the streamfunction (fig. 4.9). Especially in the mid-latitudes, the PalVeg Tortonian run represents an intensification of the wind speed (fig. 4.8). The variation of the horizontal wind field at 200 hPa (fig. 4.8b) together with the meridional cross-section of the zonal wind (fig. 4.5a) indicates not only a poleward shift of the vertical circulation but also an intensification of the jet streams in the PalVeg Tortonian run. Thus, a shift of the planetary wave, which is attributed to the palaeovegetation, is observed in the PalVeg Tortonian run.

The above described altered circulation patterns of the PalVeg Tortonian run with respect to the Standard Tortonian run refer to the annual average situation. The surface pressure (fig. 4.6b,c) and the near-surface wind field (fig. 4.7b,c) also demonstrate seasonal variations. As compared to the Standard Tortonian run, changes of the surface pressure in the PalVeg Tortonian run (fig. 4.6b,c) are more pronounced during winter than during summer of each hemisphere, respectively. On the Southern Hemisphere, the summerly (DJF) variations of the surface pressure are of low significance in the PalVeg Tortonian run (fig. 4.6c). Consequently, the summerly horizontal wind field (fig. 4.7c) is almost unaffected in the PalVeg Tortonian run. During summer, the PalVeg Tortonian run still demonstrates a slightly increased zonal mass flow as compared to the Standard Tortonian run. During southern winter (JJA), the subtropical high and the mid-latitudes low are more intense (fig. 4.6b) in the PalVeg Tortonian run. On the Southern Hemisphere, this generates stronger winterly westerlies and trade winds as compared to the Standard Tortonian simulation. Thus, the zonal mass flow of the Southern Hemisphere intensifies primarily during southern winter in the PalVeg Tortonian run.

On the Northern Hemisphere, the varied surface pressure pattern (fig. 4.6b,c) is more pronounced during winter than during summer in the PalVeg Tortonian run as compared to the Standard Tortonian run. Significant pressure variations in the PalVeg Tortonian run are seen during northern summer (fig. 4.6b). Over the tropical Atlantic Ocean, a higher pressure occurs during summer as compared to the Standard Tortonian simulation. Due to the increased

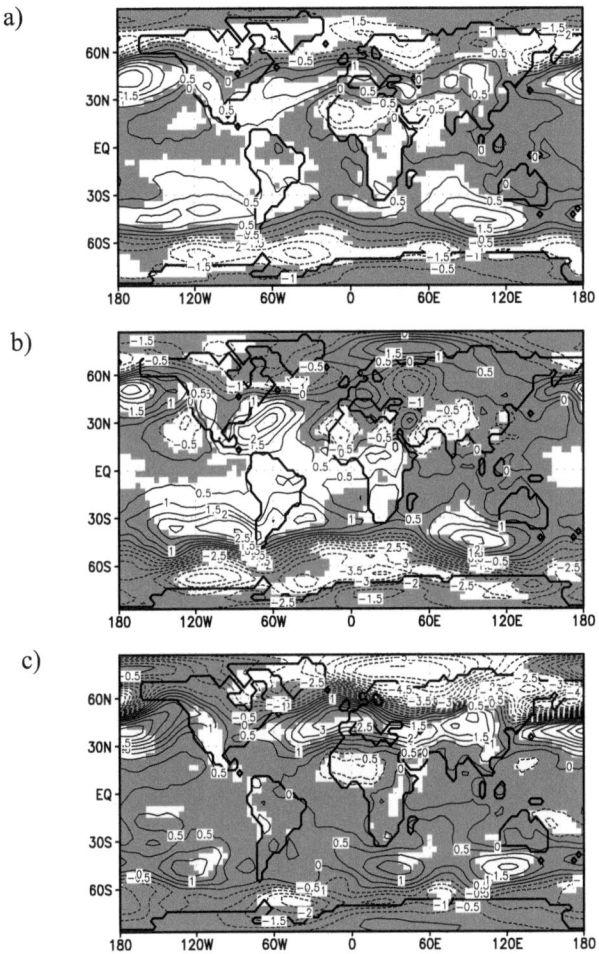

Figure 4.6: *The average surface pressure difference [hPa] between the PalVeg Tortonian and the Standard Tortonian run for a) the annual average, b) June-July-August (JJA), and c) December-January-February (DJF). The contour intervals are 0.5 hPa. Shaded areas represent non-significant anomalies with a local Student's t-test ($p = 0.05$).*

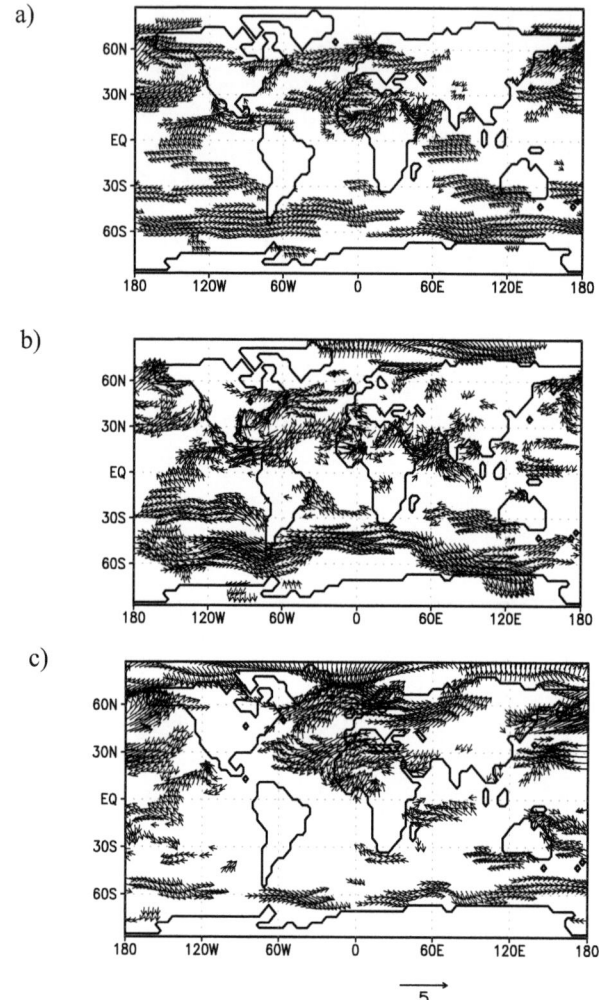

Figure 4.7: *The difference of the horizontal wind [m/s] at 1000 hPa between the PalVeg Tortonian and the Standard Tortonian run for a) the annual average, b) June-July-August (JJA), and c) December-January-February (DJF). Differences of less than 0.5 m/s for a), and 0.75 m/s for b) and c) are not shown. The reference arrow is 5 m/s.*

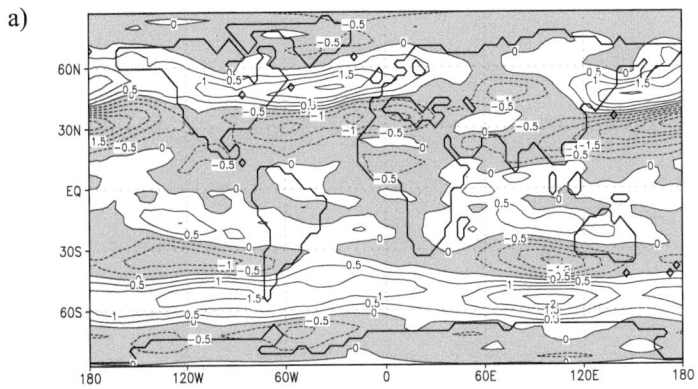

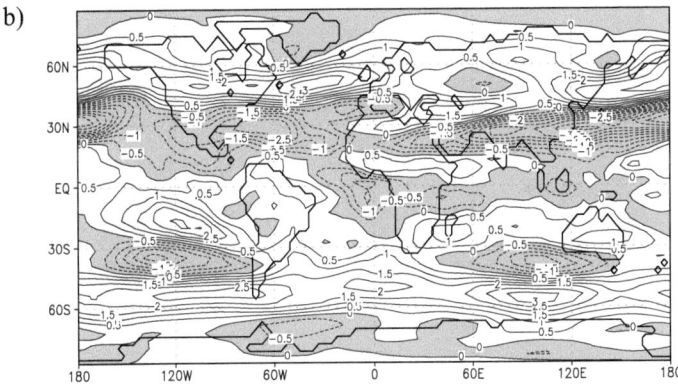

Figure 4.8: *The annual average difference of the wind speed [m/s] between the PalVeg Tortonian and the Standard Tortonian run a) at 500 hPa and b) at 200 hPa. The contour intervals are 0.5 m/s. Negative values are grey shaded.*

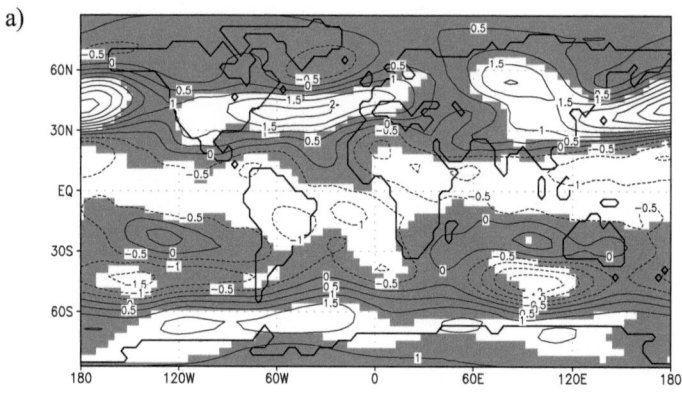

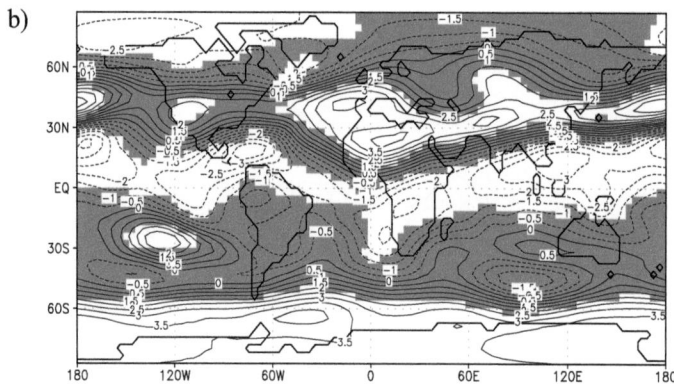

Figure 4.9: *The annual average difference of the streamfunction [10^{-6} m²/s] between the PalVeg Tortonian and the Standard Tortonian run at a) 500 hPa and at b) 200 hPa. The contour intervals are 0.5×10^{-6} m²/s. Shaded areas represent non-significant anomalies with a local Student's t-test (p = 0.05).*

surface pressure there, the meridional exchange of air masses intensifies in the in the PalVeg Tortonian simulation (fig. 4.7b). On the one hand, warmer (cf. fig. 4.2) and wetter (cf. fig. 4.3) air is transported from the lower towards higher latitudes in the PalVeg simulation. On the other hand, relatively cooler and drier air moves towards the tropical/subtropical latitudes as compared to the Standard Tortonian run. For the North Pacific, this summerly intensified exchange of air masses is also observed in the PalVeg simulation (fig. 4.6b and 4.7b). This contributes to a strengthened moisture transport from the low to the mid-latitudes in the PalVeg Tortonian run as compared to the Standard Tortonian run (cf. sec. 4.2.3). This intensified transport also leads to a reduction of the meridional temperature gradient (cf. sec. 4.2.2) as compared to the Standard Tortonian run.

During winter, the surface pressure (fig. 4.6c) indicates a shift of the vertical circulation cells on the Northern Hemisphere: The mid-latitudes get under higher pressure in the PalVeg Tortonian run, whereas the surface pressure of polar regions is reduced as compared to the Standard Tortonian run. The Northern Hemisphere's near-surface wind field (fig. 4.7c) consequently indicates a strengthened winterly zonal wind component in the PalVeg Tortonian run. During winter, parts of Northern Europe are provided with warmer and more humid air masses from the Atlantic Ocean in the PalVeg simulation (cf. sec. 4.2.3). On the Northern Hemisphere, the stronger meridional component (fig. 4.7c) demonstrates a strengthened winterly northward flow in the PalVeg Tortonian simulation. Particularly during northern winter, the palaeovegetation thus compensates for weaker palaeoceanic heat transport. The meridional temperature gradient during winter is consequently more reduced than during summer in the PalVeg Tortonian run (cf. fig. 4.1).

4.2.5 Regional atmospheric circulation patterns

The Asian monsoon

For the Himalayas, a summerly warming in the PalVeg Tortonian run was previously mentioned (cf. sec. 4.2.3). This warming causes a lower summerly surface pressure over the Himalayas as compared to the Standard Tortonian run (fig. 4.6b). During summer, the horizontal near-surface wind field (fig. 4.10) indicates a more south-west flow of the PalVeg Tortonian run as compared to the Standard Tortonian run. Thus, the palaeovegetation causes a strengthening of the Asian summer monsoon as the thermal contrast between land and ocean is higher in the PalVeg Tortonian run than in the Standard Tortonian run. The zonal averages of the summerly zonal and meridional wind components (fig. 4.11a,b) supports the intensified Asian summer monsoon in the PalVeg Tortonian run. During summer, the zonally averaged zonal wind (fig. 4.11a) indicates a northward displacement of the jet stream in the PalVeg Tortonian run.

As previously mentioned (cf. sec. 4.2.3), the warming of the Himalayas is weaker during winter than during summer in the PalVeg Tortonian run. For the Himalayas and its northern flank, the winterly surface pressure (fig. 4.6c) increases in the PalVeg simulation. During winter, the zonal average wind field (fig. 4.11) demonstrates a northward shift of the subsiding branch of the Hadley cell of about one grid cell as compared to the Standard Tortonian run. For the Indian subcontinent, the winterly near-surface wind pattern (fig. 4.10) does not indicate significant differences between both Tortonian runs. The Asian winter monsoon remains unaffected in the PalVeg Tortonian run as compared to the Standard Tortonian run.

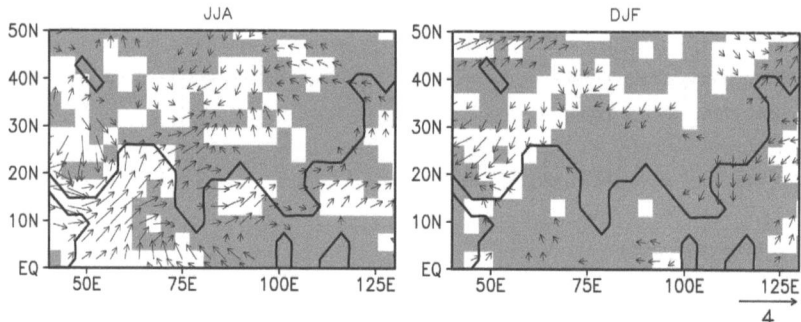

Figure 4.10: *The difference of the horizontal wind at 1000 hPa [m/s] between the PalVeg Tortonian run and the Standard Tortonian run shown for Asia (EQ to 50°N and 40°E to 130°E). The left diagram represents months June, July and August (JJA), the right diagram the months December, January and February (DJF), respectively. Values of less than 0.5m/s are not shown and the reference arrow represents 4 m/s. Shaded areas represent non-significant anomalies with a local Student's t-test ($p = 0.05$).*

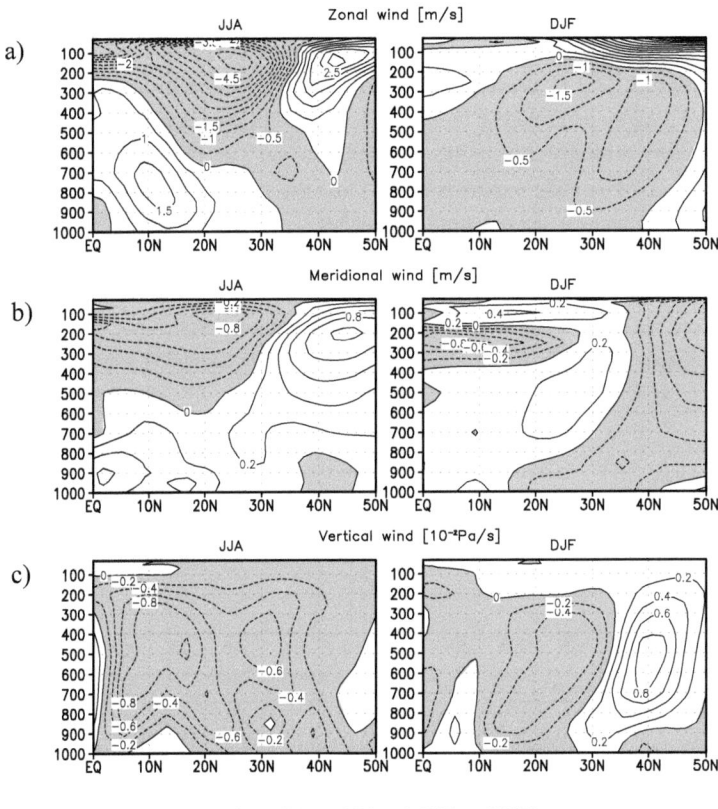

Asia (EQ to 50°N and 40°E to 130°E)

Figure 4.11: *The zonal average difference between the PalVeg Tortonian run and the Standard Tortonian run for a) the zonal wind [m/s], b) the meridional wind [m/s] and the vertical wind [10⁻²Pa/s] for Asia (EQ to 50°N and 40°E to 130°E) with respect to the height (in pressure coordinates). The left diagrams represent the months June, July and August (JJA), the right diagrams the months December, January and February (DJF), respectively. The contour intervals are 0.5 m/s for a), 0.2 m/s for b) and 0.2×10⁻² Pa/s for c), respectively. Negative values are grey shaded.*

Northern Africa and the Mediterranean

For North Africa, a warm grass and savanna vegetation replaces the modern Sahara desert (cf. fig. 3.2) in the PalVeg Tortonian run. This Tortonian vegetation has a lower albedo than the modern one (cf. tab4.1). Because of the albedo effect, the summerly solar radiation input at the surface is higher in the PalVeg Tortonian run (fig. 4.12b). During summer, this additional amount of energy as compared to the Standard Tortonian run is primarily converted into latent heat (fig. 4.12a). More water vapour is brought into the atmosphere during summer in the PalVeg Tortonian run. On the one hand, the higher summerly evapotranspiration leads to decreasing summer temperatures (–2 °C) in the PalVeg Tortonian run (fig. 4.13a). On the other hand, the summerly rainfall increases (+100 mm/mon) as compared to the Standard Tortonian run (fig. 4.13b). At around 20°N, the heat flux increases during summer in the PalVeg Tortonian simulation (fig. 4.12a), which generates an additional upward movement

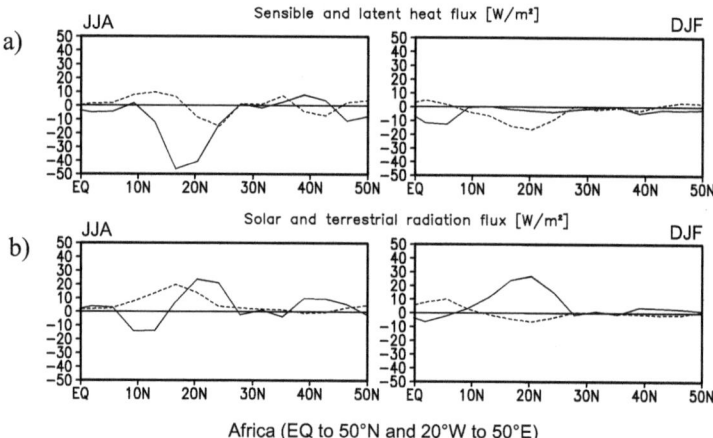

Africa (EQ to 50°N and 20°W to 50°E)

Figure 4.12: *The zonal average difference between the PalVeg Tortonian run and the Standard Tortonian run for a) the latent (**solid**) and sensible (**dotted**) heat fluxes at the surface [W/m²] and b) the solar (**solid**) and terrestrial (**dotted**) radiation fluxes at the surface [W/m²] for North Africa (EQ to 50°N and 20°W to 50°E). The left side represents the months June, July and August (JJA), the right side the months December, January and February (DJF), respectively. Negative anomalies represent an increased heat flux from the surface to the atmosphere (a) and a decreased radiation flux from the atmosphere to the surface (b).*

in the PalVeg Tortonian run (fig. 4.14c). This indicates a summerly shift of the subsiding branch of the Hadley cell towards the north (about one grid cell) as compared to the Standard Tortonian run. The summerly vertical wind anomalies (fig. 4.14c) represent a weakening of the Hadley circulation in the PalVeg Tortonian run. During summer, the atmospheric circulation (fig. 4.14) and the precipitation (fig. 4.13b) patterns indicate a strengthened African summer monsoon in the PalVeg Tortonian run.

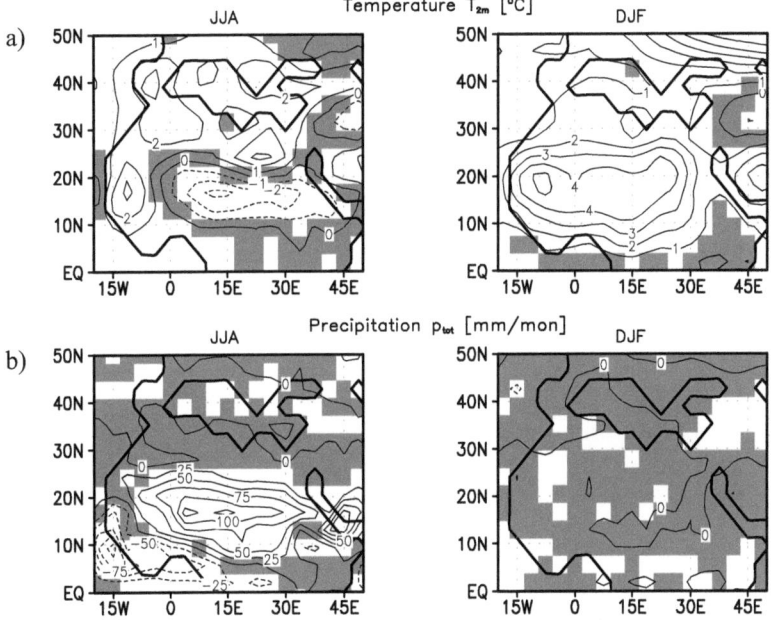

Figure 4.13: *The difference between the PalVeg Tortonian run and the Standard Tortonian run for a) the 2m-temperature [°C] and b) the total precipitation [mm/mon] for North Africa (EQ to 50°N and 20°W to 50°E). The left side represents the months June, July and August (JJA), the right side the months December, January and February (DJF), respectively. The contour intervals are 1 °C for a), and 25 mm/mon for b). Shaded areas represent non-significant anomalies with a local Student's t-test ($p = 0.05$).*

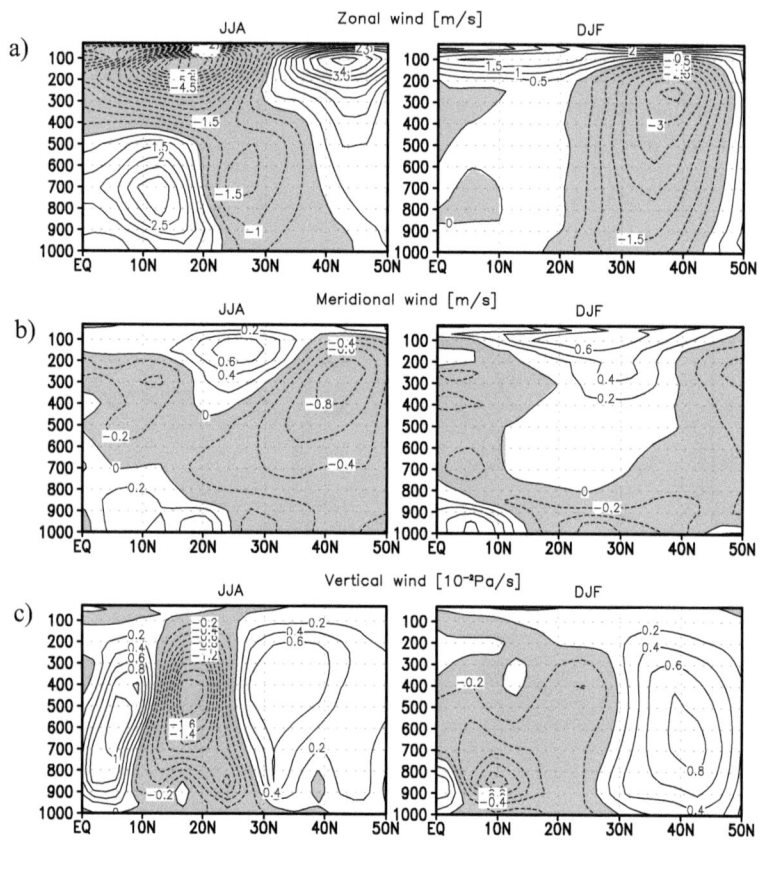

Figure 4.14: *The zonal average difference between the PalVeg Tortonian run and the Standard Tortonian run for a) the zonal wind [m/s], b) the meridional wind [m/s] and c) the vertical wind [10^{-2}Pa/s] for North Africa (EQ to 50°N and 20°W to 50°E) with respect to the height (in pressure coordinates). The left side represents the months June, July and August (JJA), the right side the months December, January and February (DJF), respectively. The contour intervals are 0.5 m/s for a), 0.2 m/s for b) and 0.2×10^{-2} Pa/s for c), respectively. Negative values are shown grey shaded.*

During winter in North Africa, the solar radiation flux at the surface increases (fig. 4.12b) in the PalVeg Tortonian run. This is attributed to the albedo effect. Contrarily to the summer situation, the winterly latent heat flux in North Africa (fig. 4.12a) is almost unaffected as compared to the Standard Tortonian run. During winter, primarily the increased sensible heat flux (fig. 4.12a) transports heat into the atmosphere in the PalVeg simulation. Accordingly fig. 4.13a shows that the North African winter temperatures rise (+4 °C) as compared to the Standard Tortonian run. The winterly rainfall does not demonstrate significant differences between both Tortonian runs (fig. 4.13b).

The vertical wind during winter (fig. 4.14c) demonstrates that the subsiding branch of the Hadley cell over North Africa extends further northward in the PalVeg Tortonian run. Corresponding to about two grid points, the winterly Hadley cell over North Africa enlarges for more than 700km towards the Mediterranean in the PalVeg Tortonian run as compared to the Standard Tortonian run. During winter in North Africa, the zonal averages of the meridional and vertical wind components (fig. 4.14b,c) represent a weakening of the Hadley circulation in the PalVeg Tortonian run. As the atmospheric circulation patterns for North Africa are affected particularly during winter in the PalVeg Tortonian run, implications on the European climate during winter are expected.

Europe and the North Atlantic storm tracks

The surface pressure during winter (fig. 4.6c) indicates an increased winterly Azores high in the PalVeg simulation. During winter, the Iceland low decreases as compared to the Standard Tortonian run. Thus, the winterly North Atlantic Oscillation (NAO) index is higher in the PalVeg Tortonian run. For the North Atlantic and European region during winter, the winterly horizontal wind patterns during winter (fig. 4.7c) demonstrate a more north-eastward flow during winter in the PalVeg Tortonian run. This indicates an intensified cyclone activity as compared to the Standard Tortonian run.

To figure out the storm activity during winter, the standard deviation of the 500 hPa geopotential of both Tortonian runs is band-pass filtered (fig. 4.15a,b) using a baroclinic filter (BLACKMON, 1976). In the mid-latitudes of both hemispheres during winter, the resulting standard deviation of the 500 hPa geopotential demonstrates an increased winterly storm activity in the PalVeg simulation (fig. 4.15a). This is in accordance to the above mentioned higher NAO index in the PalVeg simulation. For the North Atlantic Ocean and Europe during winter, the time filtered deviations of the geopotential field (fig. 4.15b) indicate that the winterly storm tracks shift northward as compared to the Standard Tortonian run. For this region, the band-pass filtered standard deviations of the geopotential field (fig. 4.15b) represent a more frequent storm activity during winter in the PalVeg Tortonian run. During winter, particularly south of Greenland over the North Atlantic Ocean and Northern Europe demonstrate positive geopotential anomalies (fig. 4.15b) as compared to the Standard Tortonian run. For these regions, this patterns correlates with increases in precipitation (+20 mm/mon) during winter in the PalVeg Tortonian run (fig. 4.15c).

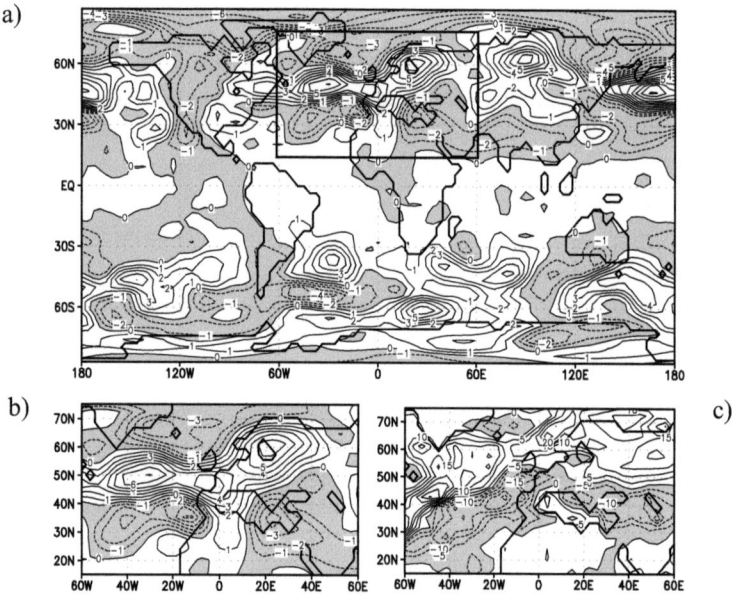

Figure 4.15: *The difference between the PalVeg Tortonian run and the Standard Tortonian run during winter (DJF) for a) the band-pass filtered standard deviation of the 500 hPa geopotential [gpm], b) the standard deviation of the band-pass filtered standard deviation of the 500 HPa geopotential [gpm] for the North Atlantic and Europe* (15°N to 75°N and 60°W to 60°E) *and c) the precipitation rate [mm/mon] for the North Atlantic and Europe. The contour intervals are 5 gpm for a), 1 gpm for b) and 5 mm/mon for c), respectively. Negative values are shown grey shaded, respectively.*

The horizontal wind (fig. 4.7c) and precipitation patterns (fig. 4.15c) during winter indicate an increased winterly moisture transport from the subtropical Atlantic Ocean to the North Atlantic Ocean and from the North Atlantic Ocean towards Europe in the PalVeg Tortonian run. In the mid-latitudes, the increased winterly storm activity thus contributes to a more efficient heat and moisture transport towards the higher latitudes as compared to the Standard Tortonian run. During winter, this more efficient heat transport leads to a reduction of the pole-to-equator temperature gradient particularly in the mid- and high latitudes (fig. 4.1) in the PalVeg Tortonian run (cf. sec. 4.2.2).

4.3 The PalVeg Tortonian run compared to the Recent Control run

The effects of palaeovegetation were demonstrated from the comparison of the PalVeg Tortonian with the Standard Tortonian run. To see the differene between the Tortonian climate and the Recent situation, the PalVeg Tortonian run is compared to the Recent Control experiment. This comparison concentrates on the temperature (sec. 4.3.1) and precipitation and

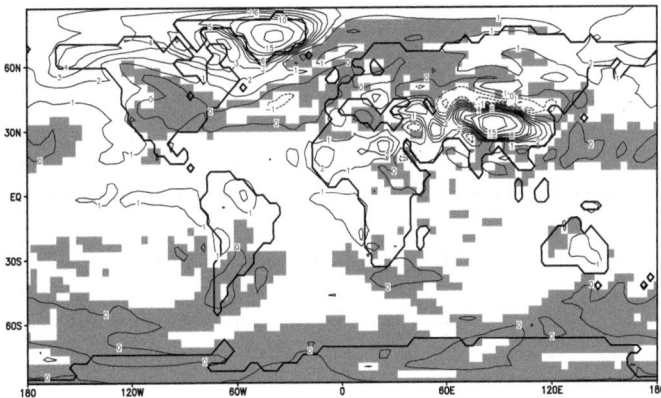

Figure 4.16: *The mean annual 2m-temperature anomalies [°C] between the PalVeg Tortonian run and the Recent Control run. Shaded areas represent, nonsignificant' anomalies with a local Student's t-test (p = 0.05).*

evapotranspiration patterns (sec. 4.3.2). Within a master thesis, which was supervised within the context of the present PhD thesis, ENGVALL (2003) focuses in detail on the description of differences between the Tortonian and the present-day's water cycle. The results of this master thesis (ENGVALL, 2003) are also briefly summarised in the following.

4.3.1 The regional temperature patterns

Fig. 4.16 shows the mean annual temperature differences between the PalVeg Tortonian run and the Recent Control experiment. For Greenland and the Himalayas, the highest increase rates of more than +15 °C are indicated as compared to today (fig. 4.16). This is in accordance to results of the Standard Tortonian run (cf. fig. 2.2). In the lower latitudes, modest warming rates (+0.5 °C to +1 °C) occur in the PalVeg Tortonian run. For the mid-latitudes, the mean annual temperatures are almost equal or slightly lower than today (–0.5 °C) in the PalVeg Tortonian run. Thus, the palaeovegetation in the PalVeg Tortonian run partly compensates for the cooling effect of the weaker-than-today ocean heat transport in Standard Tortonian run (cf. fig. 2.2). For the Northern flank of the Himalayas, cooler conditions (–2 °C) than today are observed from the PalVeg Tortonian run (fig. 4.16). During the Tortonian, parts of Southern Europe and North Africa are warmer by +0.5 °C to +1 °C as compared to the Recent Control simulation. In the high latitudes, the PalVeg Tortonian run represents generally warmer conditions than today (fig. 4.16). Particularly for Siberia and Alaska, the PalVeg Tortonian simulation demonstrates a difference of +2 °C to +4 °C with respect to present-day's conditions.

		Global			Northern Hemisphere			Southern Hemisphere		
		Total	Sea	Land	Total	Sea	Land	Total	Sea	Land
Recent Control run	p_{tot}	1019	1123	754	991	1183	685	1047	1079	902
	E	1023	1223	516	969	1290	461	1077	1173	635
PalVeg Tortonian run	p_{tot}	1046	1126	843	1023	1174	783	1069	1090	972
	E	1050	1234	585	1003	1297	539	1097	1187	684
PalVeg Torton – Control	Δp_{tot}	+27	+3	+89	+32	–9	+98	+22	+11	+70
	ΔE	+27	+11	+69	+34	+7	+78	+20	+14	+49
	$\Delta p_{tot} - \Delta E$	±0	–8	+20	–2	–16	+20	+2	–3	+21

Table 4.5: *The annual average rates [mm/a] of the precipitation, p_{tot}, the evapotranspiration, E, for the Recent Control run, the PalVeg Tortonian run and the differences between them. The annual averages are shown globally, for both hemispheres and split into land and sea, respectively.*

4.3.2 The global average precipitation and evapotranspiration

For the Tortonian, it is indicated that the global average precipitation (p_{tot}) and evapotranspiration (E) increase by +27 mm/a as compared to the Recent Control run (tab. 4.5). Over ocean surfaces, the global precipitation (+3 mm/a) and evapotranspiration (+11 mm/a) in the PalVeg Tortonian simulation are less affected. Over continental areas, higher increase rates (Δp_{tot} = +89 mm/a and ΔE = +69 mm/a) are observed from the PalVeg Tortonian run. On the global scale, the Tortonian conditions over land surfaces differ more from those over the oceans.

For the Tortonian, the net water balance, the precipitation minus the evapotranspiration ($\Delta p_{tot} - \Delta E$), is negative over the oceans (–8 mm/a) as compared to nowadays (tab. 4.5). Land surfaces demonstrate more humid conditions than today (+20 mm/a). Thus, the transport of water from the oceans towards land surfaces globally increases in the Tortonian run with

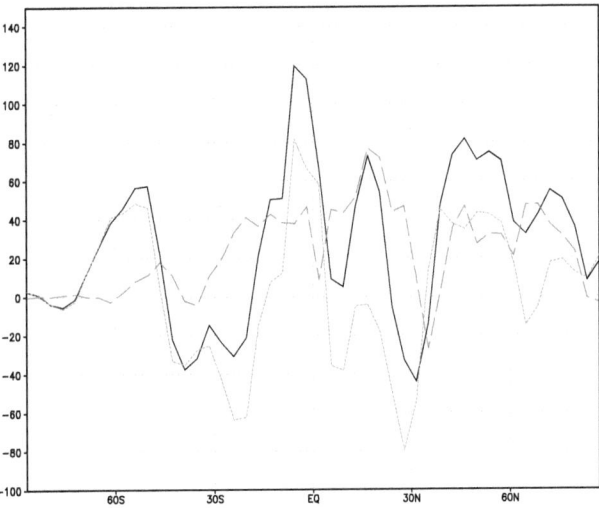

Figure 4.17: *The zonal average difference rates [mm/a] of the total precipitation (**black solid**), the evapotranspiration (**dark grey dashed**) and the precipitation minus the evapotranspiration (**light grey dotted**) between the PalVeg Tortonian run and the Recent Control run.*

respect to the Control run. Due to the different landmass distribution, variations on the Northern Hemisphere are more pronounced than those on the Southern.

4.3.3 The zonal average precipitation and evapotranspiration patterns

The global precipitation and evapotranspiration patterns increase in the PalVeg Tortonian run. The zonal averages of both, precipitation and evapotranspiration, reveal more sophisticated patterns (fig. 4.17). In the equatorial region, the zonal average precipitation increases (+120 mm/a) as compared to the Control experiment. For the subtropics, the Tortonian rainfall is lower than today (–40 mm/a). Further towards the mid- and high latitudes, the Tortonian precipitation is higher than today (fig. 4.17). There, the Northern Hemisphere demonstrates higher increases (+80 mm/a) than the Southern (+60 mm/a). In higher latitudes, the increases in precipitation in the PalVeg Tortonian run are relatively higher than those in the low latitudes.

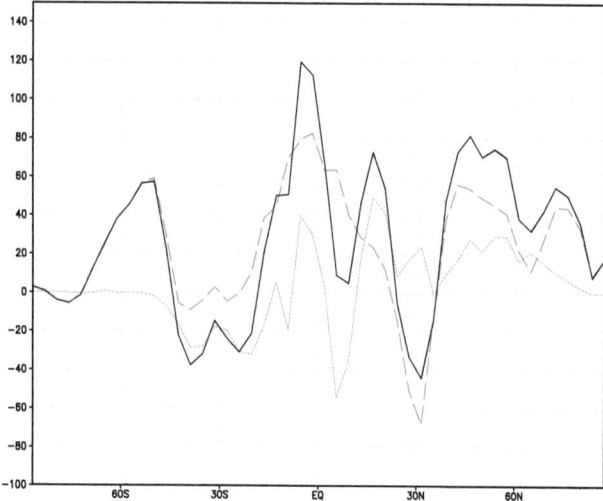

Figure 4.18: *The zonal average difference rates [mm/a] of the total (**black solid**), the large-scale (**dark grey dashed**) and the convective (**light grey dotted**) precipitation between the PalVeg Tortonian run and the Recent Control run.*

Fig. 4.17 shows that the Tortonian evapotranspiration rate generally increases as compared to nowadays. These differences in the evapotranspiration have a smaller amplitude than those of the precipitation. At around 40°N, a reduced evapotranspiration rate (–25 mm/a) is indicated from the PalVeg Tortonian run. In tropical regions between 30°S and 30°N, the evapotranspiration is higher by +40 mm/a to +75 mm/a in the Tortonian run as compared to the present-day's Control run (fig. 4.17). In mid- and high latitudes, the increases are about +20 mm/a to +40 mm/a as compared to nowadays. As compared to the absolute evapotranspiration rates, these increases in the higher latitudes are relatively higher than in the tropics. This is similar to the precipitation pattern (cf. above). In the tropical and mid- to high latitudes, the differences between the precipitation and the evapotranspiration (fig. 4.17) indicate a water excess in the PalVeg Tortonian run. The subtropical and lower mid-latitudes demonstrate a deficit as compared to the Control run.

In the tropical and higher latitudes, the large-scale (advective) precipitation (fig. 4.18) increases as compared to today. The more humid Tortonian conditions in the tropical and higher latitudes (fig. 4.17) can be attributed to an increased moisture transport as compared to nowadays. During the Tortonian, the atmospheric heat flux is thus more important than today. This is similar to a more efficient northward energy transport contributing to a shallower-than-present meridional temperature gradient (cf. sec. 4.2.2) in the PalVeg Tortonian run.

On the Northern Hemisphere, the convective precipitation generally increases in the PalVeg Tortonian simulation (fig. 4.18). Primarily over land surfaces, the convective precipitation is higher as today (ENGVALL, 2003) because land surfaces provide a larger amount of water (cf. tab. 4.5) during the Tortonian.

4.3.4 The regional precipitation and evapotranspiration patterns

Over continents, the annual precipitation (fig. 4.19a) increases in the PalVeg Tortonian run. For North Africa, the precipitation rate is strongest enhanced (+500 mm/a) in the PalVeg Tortonian run as compared to the present-day simulation. Mostly, this high increase is attributed to the palaeovegetation (+400 mm/a; cf. fig. 4.3a). For Greenland, the Tortonian precipitation is higher (+200 mm/a) than nowadays (fig. 4.19a), but the contribution of the palaeovegetation (+50 mm/a; cf. fig. 4.3a) is less than for North Africa.

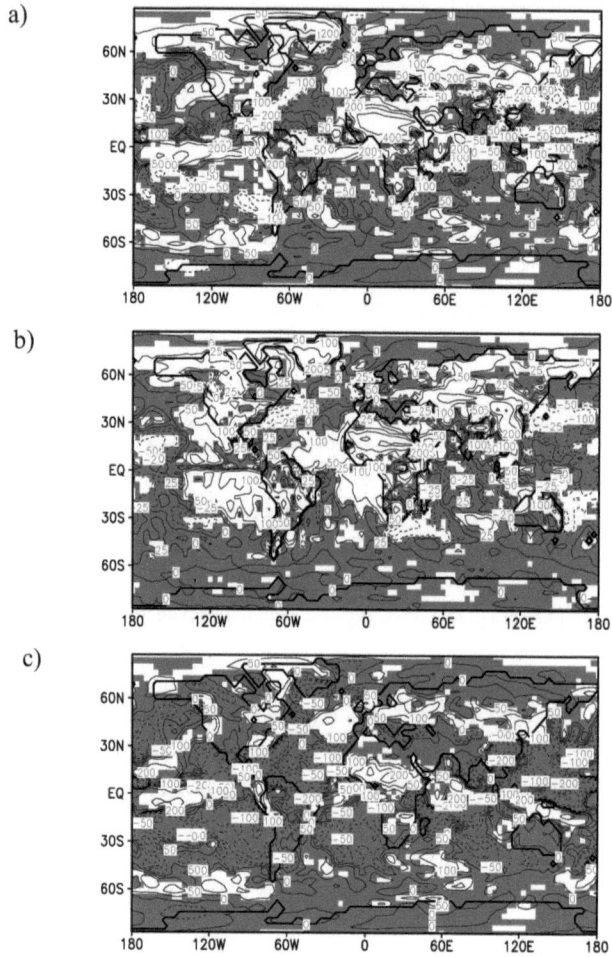

Figure 4.19: *The anomalies between the PalVeg Tortonian run and the Recent Control run for a) the annual precipitation [mm/a], b) the annual evapotranspiration [mm/a] and c) the differences between the annual precipitation and evapotranspiration [mm/a]. Shaded areas represent non-significant anomalies with a local Student's t-test (p = 0.05).*

For the oceans, the precipitation generally decreases (–200 mm/a) in the PalVeg Tortonian run as compared to the Recent Control run (fig. 4.19a). Over oceans of the southern mid- and high latitudes, the precipitation of the PalVeg Tortonian run is almost the same as for the Control run. Tropical ocean areas indicate more humid (+200 mm/a) conditions than today. The North Atlantic west of Europe indicates an increase in precipitation of +100 mm/a in the PalVeg Tortonian run (fig. 4.19a). From the subtropical Atlantic Ocean towards the North Atlantic Ocean, the transport of moisture increases in the PalVeg Tortonian run as more arid conditions are demonstrated in the low latitudes and more humid conditions in the high latitudes as compared to today.

Regarding the annual evapotranspiration (fig. 4.19b), an increase is observed for the continents in the PalVeg Tortonian run. Mostly, the higher-than-present evapotranspiration is caused by the palaeovegetation (cf. fig. 4.3b). The larger amount of water vapour in the atmosphere then increases the precipitation as compared to today (cf. above). For North Africa, +300 mm/a are evaporated into the atmosphere as compared to today. Almost completely, this increase is attributed to the palaeovegetation (cf. fig. 4.3b). For ocean surfaces, the evapotranspiration patterns of the PalVeg Tortonian run and the Control run do not differ much. Ocean surfaces of the equatorial region demonstrate increases in evapotranspiration (+100 mm/a) as compared to the present-day's situation (fig. 4.19b). For the regions of the Gulf Stream and the Kuroshio current, evapotranspiration rates are lower (–100 mm/a to –200 mm/a) than present in the PalVeg Tortonian run. This reduction partly contributes to reduce the precipitation as compared to the Control run.

In equatorial latitudes as well as in the high latitudes, the net balance of precipitation minus evapotranspiration (fig. 4.19c) represents more humid (+100 mm/a to +300 mm/a) conditions in the PalVeg Tortonian run. For ocean surfaces of the mid-latitudes, more arid (–100 mm/a to –200 mm/a) conditions than today are observed (fig. 4.19c). Thus, this indicates the strengthened transport of moisture from the lower towards the high latitudes in the Tortonian simulation.

4.4 Discussion

In the following subsection 4.4.1, the weak points of the ECHAM model and insufficiencies in the model setup for the Tortonian are discussed. In oder to verify the PalVeg Tortonian run, its results are compared to other modelling studies (sec. 4.4.2). As Miocene modelling studies are quite rare in literature, various model studies for periods such as the Holocene or the Cretaceous are also considered. A comparison to proxy data is not presented in this chapter but will follow later on (sec. 5).

4.4.1 Weak points of the model and of the setup of the PalVeg Tortonian run

It should be noticed that models themselves include some systematic weak points. The vegetation module of ECHAM4 is such an example of a simplifying parameterisation (DKRZ MODELLBETREUUNGSGRUPPE, 1994). In ECHAM4, the biosphere is simply treated as a non-dynamical system. If the biosphere is represented as a dynamical parameter, SCHNITZLER ET AL. (2001) demonstrate some more realistic model results (e.g., the variability of rainfalls). Even though some weak points are obvious, ECHAM demonstrates a quite good performance for multiple present-day and Quaternary palaeostudies (LATIF & NEELIN, 1994; LORENZ ET AL., 1996; MONTOYA ET AL., 1998). However, the mentioned Tortonian simulations, the Standard Tortonian run (STEPPUHN ET AL., 2006), the $2\times CO_2$ Tortonian run (STEPPUHN ET AL., 2007) and the PalVeg Tortonian run (cf. sec. 4) represent the first approaches of applying the ECHAM model to the Tertiary.

Amongst these systematic difficulties of the model itself, some weak points and uncertainties in the setup regarding the vegetation in the PalVeg Tortonian run have to be mentioned, as it is the only changed parameter compared to the previous Standard Tortonian run (cf. sec. 4.1). Proxy data of a single (small) location are assumed to be representative for a whole grid cell. This assumption is needed but not satisfied necessarily at each location. Fossil data include some uncertainties (UHL & BRUCH, pers. comm.). In addition, the calculated palaeovegetation distribution using a biome model (cf. sec. 3) is not fully correct since using the slightly unrealistic data of the Standard Tortonian run (STEPPUHN ET AL., 2006). Moreover, the Prentice biome model itself produces some inaccuracies (PRENTICE ET AL., 1992; CLAUSSEN, 1993). Despite the weak points, it can be assumed that the reconstructed Tortonian vegetation is quite realistic. At last for the purpose of global climate modelling, it also includes a larger number of biome classes in a slightly higher spatial resolution than previous global reconstructions (WOLFE, 1985) used for model simulations of the Early Miocene (DUTTON & BARRON, 1997).

In order to consider the Tortonian vegetation, surface parameters applying to vegetation are adapted in the palVeg Tortonian run (cf. sec. 4.1). Surface roughness lengths (z_0) remain unchanged as compared to the Standard Tortonian run. Comparing the Recent and the Tortonian vegetation, some differences are apparent (cf. sec. 3). According to CLAUSSEN (1994), variations of the vegetation roughness length ($z_{0,veg}$) of different biomes are in the order of a few decimetres when replacing the modern with the Tortonian vegetation. The orography also influences the surface roughness ($z_{0,oro}$). It is known that the height of the mountain ranges generally increases since the Late Miocene (cf. sec. 2.2.1 and sec. 4.1). However, the reconstruction of the palaeorography includes some uncertainties for the elevation (KUHLEMANN, pers. comm.) and, therefore, uncertainties for $z_{0,oro}$. CLAUSSEN (1994) calculates the surface roughness length from the following equation:

$$z_0 = \sqrt{z_{0,oro}^2 + z_{0,veg}^2} \,.$$

Particularly in mountaineous regions, this equation indicates that uncertainties (δ) in the roughness lengths of the orography can be of more importance than those of the vegetation ($\delta z_{0,oro} \gg \delta z_{0,veg}$). Accordingly CLAUSSEN (1994) emphasises that variations of the surface roughness z_0 due to changes in vegetation are rather small in mountaineous regions. Thus, an unrealistic representation of the Tortonian climate in the PalVeg Tortonian run can be rather more attributed to uncertainties in the palaeorography than to the unchanged vegetation roughness lengths.

In addition, DUTTON & BARRON (1996) mention that effects of varying roughness lengths due to an altered vegetation are of minor importance as compared to the albedo-effect and the effect on the water cycle. During the Earth's history, particularly the water cycle plays an important role in the climate system (BARRON ET AL., 1989). For the ECHAM4 model, KLEIDON & HEIMANN (2000) demonstrate a sensitivity of modelling results to the rooting depth and, hence, the maximum available soil water capacity, which is adapted for the PalVeg Tortonian run (cf. sec. 4.1). Therefore it can be concluded that the unchanged roughness lengths of the PalVeg Tortonian run are of minor importance in relation to the uncertainties of other parameters such as the maximum available soil water capacity.

4.4.2 The comparison of the PalVeg Tortonian run with other model results

Climatic changes over ocean surfaces are observed from the PalVeg Tortonian run as compared to the Standard Tortonian run (cf. sec. 4.2). These represent teleconnection patterns, which are attributed to the palaeovegetation. The atmospheric pattern of the Pacific region in the PalVeg Tortonian run resembles a permanent El Niño (cf. sec. 4.2). This pattern is also stated for the Standard Tortonian run (STEPPUHN ET AL., 2006). However, the El Niño-phenomenon (BIGG, 1999, LATIF & NEELIN, 1994), which is caused by interactions between the atmosphere and the oceans, cannot be simulated with a simple mixed-layer ocean model. The PalVeg simulation demonstrates a higher NAO index and a response of the North Atlantic storm tracks as compared to the Standard Tortonian run (cf. sec. 4.2.5). This can be related to the El Niño-like pattern in the Pacific region. Based on Recent simulations, CARILLO ET AL. (2000) accordingly observe an effect on the storm track regimes during El Niño.

The oceans

In the North Atlantic Ocean, a higher freshwater input can be assumed due to an increased precipitation in the PalVeg Tortonian run (cf. sec. 4.2.3). The conditions over the Central Atlantic Ocean are more arid than in the Standard Tortonian run. A change in the hydrological cycle over ocean surfaces in the PalVeg Tortonian run alters the ocean salinity. Consequently, the influence of the palaeovegetation should reach the oceans and should affect the oceanic heat transport. An effect on the ocean circulation cannot be demonstrated from the PalVeg Tortonian run because of the used mixed-layer ocean model (STEPPUHN ET AL., 2006). However, it can be supposed that lower saline high latitude Atlantic Ocean water weakens the poleward oceanic heat transport, whereas higher saline conditions in the subtropical Atlantic Ocean strengthen it. A weak ocean heat transport would be consistent with ocean modelling studies, which demonstrate that an open Panama seaway, as it existed before 3Ma, causes a weaker northward heat transport in the Atlantic Ocean (MAIER-REIMER ET AL., 1990; MIKOLAJEWICZ & CROWLEY, 1997). From a Recent model study, MIKOLAJEWICZ & VOSS (2000) observe an increased freshwater influx in the North Atlantic Ocean, which is attributed to an increased atmospheric CO_2 concentration. This causes a weakening of the poleward heat transport in the ocean (MIKOLAJEWICZ & VOSS, 2000). However, the study also demonstrates that the circulation in the Atlantic Ocean is intensified due to an increased evaporation in the subtropics (cf. sec. 4.2.3), though it is not strong enough to compensate for the weaking in the high latitudes

(MIKOLAJEWICZ & VOSS, 2000). For the last glacial period, models indicate that small changes in the hydrological cycle affect the thermohaline circulation in the Atlantic Ocean, which lead to an abrupt climate change (CLARK ET AL., 2002; MANABE & STOUFFER, 1997).

The ocean circulation is not only affected by salinity effects, but also by wind stress. An altered ocean circulation can be expected when considering the different-than-today wind field in the PalVeg Tortonian run (cf. sec. 4.2.4). OGCM studies, which investigate the effects of an open Central American Isthmus during the Miocene, often use observed present-day's wind data (MAIER-REIMER ET AL., 1990; MIKOLAJEWICZ ET AL., 1993; MIKOLAJEWICZ & CROWLEY, 1997). To perform uncoupled OGCM runs, BICE ET AL. (2000) use data of simple Early Eocene (~55 Ma) to Middle Miocene (~14 Ma) AGCM simulations. According to these simulations, the development of ocean basins is more important for the ocean circulation than the surface forcing (BICE ET AL., 2000). BICE ET AL. (2000) do not use different atmospheric forcings for the same basin configuration at a specific time slice. Therefore, the effects of a different surface forcing cannot be separated from those, which are caused by basin changes (BICE ET AL., 2000). For the Recent situation, MIKOLAJEWICZ & VOSS (2000) perform a high CO_2 model simulation using a coupled ocean-atmosphere general circulation model. Attributed to a northward shift of the atmospheric wind patterns in this study, the subtropical and subpolar gyre in the Atlantic and Pacific Ocean shift northward (MIKOLAJEWICZ & VOSS, 2000). The shift of the ocean gyres under warmer-than-present conditions implies a strengthened northward ocean heat transport (MIKOLAJEWICZ & VOSS, 2000). This can also be supposed to be the case in the PalVeg Tortonian run as a northward displacement of the atmospheric wind patterns is observed (cf. sec. 4.2.4 and 4.2.5).

In the case of the PalVeg Tortonian run as compared to the Standard Tortonian run, palaeovegetation can weaken or strengthen the North Atlantic thermohaline circulation. On the one hand, salinity effects in the low latitudes and the different wind stress forcing could strengthen the poleward oceanic heat transport. On the other hand, the dilution of the North Atlantic Ocean water masses could weaken the thermohaline circulation. However, it cannot be answered, which process dominates.

The global atmospheric pattern

The change from the Recent vegetation to an appropriate palaeovegetation is noticeable on the global scale, where the PalVeg Tortonian run demonstrates a global warming of +0.9 °C as compared to the Standard Tortonian run (cf. sec. 4.2.1). This is consistent with other findings. A global average warming/cooling of ±1 °C attributed to changes in vegetation is reported from sensitivity experiments with the GENESIS model (DUTTON & BARRON, 1996). Ascribed to a more dense forest vegetation during the Early Miocene, DUTTON & BARRON (1997) demonstrate a warming of globally +2 °C, which is more pronounced on the Northern than on the Southern Hemisphere. For the Cretaceous, a vegetation-induced global warming of +2 °C is observed from OTTO-BLIESNER & UPCHURCH (1997).

The zonal average pattern

Focusing on the zonal average temperature, the PalVeg Tortonian run as compared to the Standard Tortonian simulation indicates a reduced equator-to-pole temperature difference (cf. sec. 4.2.2). STEPPUHN ET AL. (2006) observe a flattening of the meridional temperature gradient in the Standard Tortonian run with respect to the Recent Control experiment (cf. sec. 2.2). Thus, the PalVeg Tortonian run demonstrates the shallowest gradient as compared to the Standard Tortonian and the Recent Control run. For the Early Miocene (24 to 16 Ma), DUTTON & BARRON (1997) perform two simulations with the GENESIS model. The first Miocene run includes Miocene boundary conditions combined with the Recent vegetation, while the second one uses Miocene boundary conditions with an appropriate palaeovegetation (DUTTON & BARRON, 1997). Analogous to the Tortonian experiments with ECHAM4/ML (cf. sec. 2.2 and 4.2.2), DUTTON & BARRON's (1997) Miocene simulation with adapted palaeovegetation demonstrates the shallowest meridional temperature gradient. This is most evident on the Northern Hemisphere when compared to the Miocene run with Recent vegetation and the Recent Control run (DUTTON & BARRON, 1997).

Results of the PalVeg Tortonian run with ECHAM4/ML agree quite well with those of the Early Miocene GENESIS study (DUTTON & BARRON, 1997), but some differences are observed. It must be noticed that DUTTON & BARRON (1997) use surface temperatures, whereas the analysis of the PalVeg Tortonian run considers the 2m-temperatures (cf. tab. 4.3). In the following, the surface temperatures of the PalVeg Tortonian run are compared to results of DUTTON & BARRON's (1997) Early Miocene study. The GENESIS model simulates warmer conditions for the Early Miocene (T_{NH} = 22.2 °C, T_{SH} = 14.4 °C) than it is observed from the PalVeg Tortonian run with ECHAM (T_{NH} = 17.0 °C, T_{SH} = 15.8 °C). Regarding the Recent

Control experiments, the GENESIS model ($T_{global,ctrl}$ = 14.4 °C) represents cooler conditions than the ECHAM model ($T_{global,ctrl}$ = 15.7 °C). Simulations of future greenhouse scenarios with various models also reveal such differences between the models, although the boundary conditions for these experiments are the same (IPCC, 2001).

Regarding both palaeosimulations, it must be noticed that the climate successively cools during the Miocene (cf. fig. 1.1; CROWLEY, 2000; PARTRIDGE ET AL., 1995), which is attributed to changing boundary conditions. The Early Miocene run (DUTTON & BARRON, 1997) and the PalVeg Tortonian run (cf. sec. 4.1) differ in their setup (e.g. the land-sea distribution). Accordingly, some climatic differences are induced as DUTTON & BARRON (1997) use a simple Early Miocene vegetation distribution, while the PalVeg Tortonian run uses a more detailed Late Miocene palaeovegetation (cf. sec. 3). Additionally, the PalVeg Tortonian run includes a weaker-than-present ocean heat transport (cf. sec. 4.1; STEPPUHN ET AL., 2006). It can be assumed that DUTTON & BARRON's (1997) Early Miocene simulation includes an increased northward oceanic heat transport. DUTTON & BARRON (1997) do not describe the ocean setup, but this could explain the warmer Northern (ΔT_{NH} = +5.2 °C) and the cooler Southern Hemisphere (ΔT_{SH} = −1.4 °C) in the GENESIS simulation with respect to the PalVeg Tortonian run. In a simulation of the last interglacial, an increased northward oceanic heat transport also causes a cooling of the Southern Hemisphere (CROWLEY, 1992).

For the PalVeg Tortonian run, a weaker palaeoceanic heat transport is prescribed (cf. sec. 2.2.1 and 4.1). In contrats, the latent heat flux indicates a more efficient atmospheric heat transport in the PalVeg Tortonian run (cf. sec. 4.2.3 and 4.3.3). Consistently AGCM sensitivity studies with a varying ocean heat transport demonstrate that the atmospheric heat transport partly compensates for a weaker ocean heat transport (COVEY & THOMPSON (1989). In the same study, the latent heat transport is more affected than the sensible heat transport (COVEY & THOMPSON, 1989).

The high latitudes

Attributed to a strong albedo-effect, cold-continental regions such as Siberia are warmer in the PalVeg Tortonian run (cf. sec. 4.2.3), the Arctic sea ice volume decreases as compared to today (cf. sec. 4.2.1) and the Tortonian water cycle is intensified as compared to today (cf. sec. 4.2.3). Other modelling studies also report a significant albedo-effect in the high latitudes due to changes in the vegetation (DUTTON & BARRON, 1996; DUTTON & BARRON, 1997; OTTO-BLIESNER & UPCHURCH, 1997; UPCHURCH ET AL., 1998). A sensitivity study (DUTTON & BARRON, 1996) demonstrates a significant influence of vegetation on the water cycle, which is comparable to the PalVeg Tortonian run.

Asia

The Miocene monsoon is weaker than nowadays in the Standard Tortonian run (STEPPUHN ET AL., 2006). Due to the Tortonian vegetation, a strengthening of the Asian summer monsoon is observed in the PalVeg Tortonian run (cf. sec. 4.2.5). However, the Asian monsoon remains weaker than today in the PalVeg Tortonian run. Several authors discuss the Neogene uplift of the Tibetan Plateau and its influence on the Asian monsoon (FLUTEAU ET AL., 1999; RAMSTEIN ET AL., 1997). According to these studies, the Miocene Asian monsoon is weaker than today. Simulations of the Mid-Holocene demonstrate a strengthened Asian summer monsoon, which is attributed to variations in vegetation (GANOPOLSKI ET AL., 1998a).

North Africa

For North Africa, the replacement of the modern Sahara desert with warm grass vegetation induces more humid conditions during summer and a strengthened African summer monsoon in the PalVeg Tortonian run (cf. sec. 4.2). From Holocene sensitivity experiments, a self-inducing mechanism (CHARNEY, 1975) is found when greening the Sahara desert (CLAUSSEN ET AL., 1999; DE NOBLET-DUCOUDRE ET AL., 2000; GANOPOLSKI ET AL., 1998a). During the Holocene, vegetation feedback processes such as a greening of the Sahara desert tend to increase convection (CLAUSSEN ET AL., 1998), which strengthens the African summer monsoon (DOHERTY ET AL., 2000).

Europe

The PalVeg Tortonian run indicates a high winterly NAO index correlated with an intensification of the winterly storm track activity as compared to the Standard Tortonian run (cf. sec. 4.2.5). For a CO_2-induced warm Recent climate, LUNKEIT ET AL. (1998) describe variations of the storm tracks. Due to an increased CO_2 in the atmosphere, storm tracks shift eastward and the cyclone density north-eastward (SCHUBERT ET AL., 1998). Based on the analysis of Recent observation data, BLENDER ET AL. (1997) suggest a link between the variability of the winterly NAO index and moisture transport. A high NAO index is related to an increased north-eastward moisture transport (BLENDER ET AL., 1997) as consistently observed in the PalVeg Tortonian run (cf. sec. 4.2.5). As found out from present-day's data analysis (ROGERS, 1997), the winterly variability of the storm tracks correlates with changes in winter temperatures in Northern Europe. ROGERS (1997) observes a strengthening of the westerlies with an increased maritime flow during winter, which is correlated to milder winters in Europe. This is consistent to the PalVeg Tortonian run.

Generally, the PalVeg Tortonian run is largely consistent with other Miocene modelling experiments (DUTTON & BARRON, 1997; FLUTEAU ET AL., 1999; RAMSTEIN ET AL., 1997). Simulations for various periods of the Earth's history (CLAUSSEN ET AL., 1998; DUTTON & BARRON, 1997; OTTO-BLIESNER & UPCHURCH, 1997; UPCHURCH ET AL., 1998) support the effects of the Tortonian vegetation, as outlined from climatic differences between the PalVeg Tortonian run and the Standard Tortonian run. It can therefore be assumed that the appropriately considered palaeovegetation contributes to a realistic representation of the Tortonian climate in the PalVeg Tortonian run. In order to validate this, model results of the PalVeg Tortonian run are compared to proxy data in the following chapter.

5 VALIDATION OF MODEL RESULTS WITH PROXY DATA

As compared to proxy data, STEPPUHN ET AL.'s (2006) Standard Tortonian run represents some unrealistic patterns (cf. sec. 2.3.3). In order to test whether the palaeovegetation contributes to a more realistic representation of the Tortonian climate in the ECHAM model, results of the PalVeg Tortonian run are now compared with proxy data covering the whole Tortonian (11 to 7 Ma). The comparison is separated into a quantitative (sec. 5.2) and a qualitative part (5.3). The qualitative comparison is based on so-called 'soft' proxy data such as fossil mammals. Quantitative climate information is obtained from 'hard' proxy data: micro- (pollen) and macro- (leaves) botanical fossils.

5.1 Methods and data

In order to validate the PalVeg Tortonian run quantitatively, UHL & BRUCH (pers. comm.) provide Tortonian proxy data for mean annual temperature (MAT) and mean annual precipitation (MAP). On the one hand, UHL & BRUCH (pers. comm.) use the *coexistence approach* (MOSBRUGGER & UTESCHER, 1997) to specify the Tortonian climate from fossil floras. The coexistence approach (MOSBRUGGER & UTESCHER, 1997) is based on the assumption of a relationship between a fossil taxa and its nearest living relative: As the climatic environment of Recent taxa is known, for each related taxon of a fossil flora a specific minimum-maximum range for the MAT and MAP is derived. The joint-intersection of all taxa intervals results in the so-called *coexistence interval*. These coexistence intervals for MAT and MAP, respectively, are equal are smaller than each of the single intervals.

On the other hand, the compilation of Late Miocene proxy data (UHL & BRUCH, pers. comm.) includes results of other studies, which use the *leaf margin analysis* (WOLFE, 1979) and the *CLAMP method* (WOLFE, 1993). Both approaches are based on the physiognomic characters of leaves: The leaf physiognomy of plants is assumed to be an adaptation to its climatic environment (WOLFE, 1979; WOLFE, 1993). The CLAMP method (WOLFE, 1993) is an improvement of the leaf margin analysis (WOLFE, 1979). From modern plant distributions and modern climate observations, a calibration dataset is obtained, which considers the leaf physiognomy. If the percentage composition of the physiognomic characters of a fossil flora is known, average values (with an assigned standard deviation) of climatic parameters such as the mean annual temperature or the annual precipitation can be determined (WOLFE, 1993).

For the present comparison of model results and proxy data, the mean annual temperature and the annual precipitation for the Tortonian are finally derived from studies of BRUCH (1998), BRUCH ET AL. (2006), GRAHAM (1998), GREGOR & UNGER (1988), JACOBS (1999), JACOBS & DEINO (1996), MAI (1995), MARTIN (1990), MARTIN (1998), MUDIE & HELGASON (1983), PLAYFORD (1982), SACHSE & MOHR (1996), UTESCHER ET AL. (2000), WOLFE (1994a), WOLFE (1994b). Fig. 5.1 summarises the comparison of the PalVeg Tortonian run with terrestrial proxy data of each location for MAT and MAP, respectively (cf. fig. 2.2).

According to the ECHAM T30 grid, the proxy-based mean annual temperatures are additionally zonally averaged. Fig. 5.2 shows these zonal average proxy temperatures including the corresponding minimum-maximum range. Considering land surface grid points of the ECHAM model, the zonal average temperatues of the PalVeg Tortonian simulation are compared to the proxy-based zonal average temperatures. As most proxy data between 30°N and 60°N refer to Europe (cf. fig. 5.1), the European zonal average temperatures of the PalVeg Tortonian run are also plotted in fig. 5.2.

5.2 The quantitative comparison

Generally, the mean annual temperatures of the PalVeg Tortonian run and the terrestrial proxy data are consistent (fig. 5.1a). This agreement is better than for the Standard Tortonian run and proxy data (cf. fig. 2.2a). For North America, the temperatures of the PalVeg Tortonian run are still below the proxy-based estimations (fig. 5.1a). In Alaska, a difference of about −5 °C is observed between the PalVeg Tortonian run and proxy data. Further south of Alaska, the differences are smaller between the PalVeg run and proxy data. In Siberia, the PalVeg Tortonian run is much cooler (−10 °C) than proxy data suggest. A western Iceland location indicates a difference of −3 °C between the PalVeg Tortonian run and proxy data. For a more eastern part of Iceland, the PalVeg Tortonian simulation and proxy data agree. The model overestimates the MAT of China by +3 °C. For Kenya, the temperature in the Tortonian model run is only +1 °C to +2 °C higher than the proxy estimations suggest. For New Guinea, the PalVeg Tortonian run agrees with proxy data. On the one hand, the tropical regions tend to

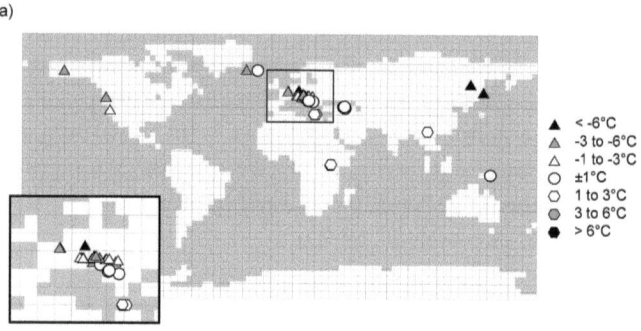

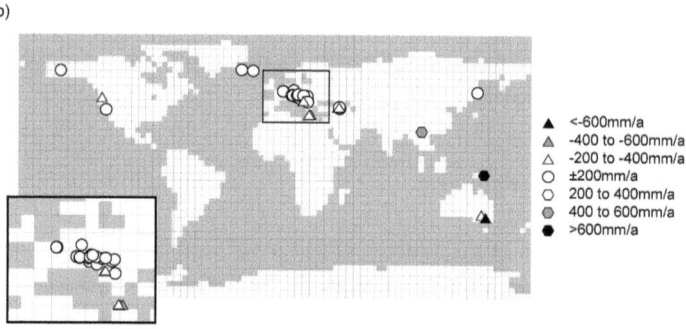

Figure 5.1: *The differences between the PalVeg Tortonian run and terrestrial proxy data for a) the mean annual temperature [°C], and b) the annual precipitation [mm/a]. The European region is shown enlarged. White circles represent consistency, triangles represent cooler or more arid conditions, and sexangles represent warmer or more humid conditions in the model simulation, respectively.*

be slightly too warm in the PalVeg Tortonian run. On the other hand, the high latitudes still remain too cool as comparted to proxy data. Thus, the meridional temperature gradient of the PalVeg Tortonian simulation is not as shallow as proxy data suggest. The discrepancies between model and proxy data are smaller in the PalVeg Tortonian run than in the Standard Tortonian run. Focusing on Europe, where most quantitative proxy data are available, the Mediterranean region such as Crete is slightly warmer (+1 °C) in the PalVeg Tortonian run than suggested by proxy data (fig. 5.1a). For Central Europe and particularly the Alps, the PalVeg Tortonian run underestimates MATs of maximally –6.7 °C as compared to proxy data. On the Balkan, the PalVeg Tortonian run agrees to the proxy estimations.

As evident from fig. 5.2, the tropics and subtropics are represented warmer in the PalVeg Tortonian run as it is suggested from proxy data. This is also demonstrated in the Standard Tortonian run. The high latitudes are cooler in both Tortonian model simulations. Thus,

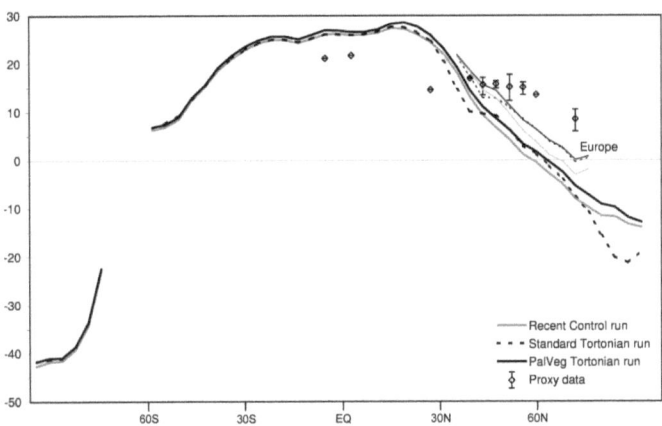

Figure 5.2: *The zonal average temperature for land surfaces of the Recent Control run (**grey solid**), the Standard Tortonian run (**black dashed**), the PalVeg Tortonian run (black solid) and terrestrial proxy data (**diamonds**) with the vertical bars indicating the minimum/maximum range. In order to obtain the zonal averages of terrestrial proxy data, the single data locations are transformed into the grid point resolution (3.75°) of the model ECHAM4/ML. For Europe, the zonal average temperatures of the model runs are shown separately.*

the meridional temperature gradient is still too steep in the PalVeg Tortonian run, but the discrepancies to proxy data are smaller than for the Standard Tortonian run. Regarding Europe, the PalVeg Tortonian run is closer to proxy data than the Standard Tortonian run (fig. 5.2). In particular from 35°N to 45°N, the agreement between the PalVeg Tortonian run and proxy data is quite good. Further northward, the PalVeg and the Standard Tortonian run do not differ much with respect to the zonal average temperatures (fig. 5.2 [cf. sec. 4.2.2]) and the deficits of the Standard Tortonian run still remain in the PalVeg Tortonian run.

Focusing on the precipitation, the PalVeg Tortonian run is globally quite consistent to proxy data (fig. 5.1b). In the high latitudes of the Northern Hemisphere, the precipitation pattern agrees with proxy data in the PalVeg Tortonian run. Just for a single location in North America the PalVeg Tortonian run underestimates (–270 mm/a) the MAP of proxy data. For the southern mid-latitudes of Australia, the climatic conditions of the PalVeg Tortonian run are drier than terrestrial proxy data suggest (–280 mm/a to –780 mm/a). Contrarily, the tropics are too humid in the model. For China, the PalVeg Tortonian run overestimates the annual precipitation with +400 mm/a as compared to proxy data. For New Guinea, the MAP in the PalVeg Tortonian run is even +2200 mm/a higher than proxy data suggest. The northern mid-latitudes of the PalVeg Tortonian simulation agree with proxy data. For a Crete location, the PalVeg Tortonian run represents drier conditions (–300 mm/a to –550 mm/a) than the proxy-based estimations (cf. Australia). Thus, the lower mid-latitudes of both hemispheres (30° to 40°) tend to be too dry in the PalVeg Tortonian run as compared to proxy data.

5.3 The qualitative comparison

Europe

In comparison to terrestrial proxy data, the PalVeg Tortonian run is more consistent to proxy data than the Standard Tortonian run (cf. sec. 5.2). For the Mediterranean region, the PalVeg Tortonian run represents some warmer and drier conditions than today (cf. sec. 4.3). For the south of the Iberian Peninsula, a temperate climate during the latest Tortonian is indicated from proxy data (BRACHERT, 1996). During the Late Miocene, the north-eastern part of Spain is a transitional region from a temperate humid to a subtropical dry climate (ALONSO-ZARZA & CALVO, 2000). For central and northern Italy during the late Tortonian, BERTINI (1994) demonstrates subtropical to warm temperate conditions, which shift to a warm-temperate climate during the Zanclean. Before and after 10.5 and 8.5 Ma, fossil small mammals from Spain indicate more arid and warmer conditions (VAN DAM & WELTJE, 1999). This is consistent with the PalVeg Tortonian run. VAN DAM & WELTJE (1999) demonstrate a more humid and cooler climate in Spain between 10.5 and 8.5 Ma. An aridification in the Mediterranean is reported for the following era of the Messinian (7 to 5.3 Ma) finally culminating in the Messinian Salinity Crisis (BENSON ET AL., 1995; BLANC, 2000).

For Central Europe, the PalVeg Tortonian run indicates a climate, which is almost as warm as today but more humid (cf. sec. 4.3). For the Lower Rhine Embayment during the Tortonian, a Cfa climate with temperatures between 15 °C to 20 °C is suggested (GEBKA ET AL., 1999). During the Late Miocene, a transition from a warm and dry Southern European climate to a wet-dry seasonal climate in Northern Europe occurs (VAN DAM & WELTJE, 1999), which tends to agree better with the PalVeg Tortonian run than with the Standard Tortonian run (cf. sec. 2).

North America

For the mid-latitudes of the North American continent, the PalVeg Tortonian run represents almost the same mean annual temperature than today but rainfalls are higher (cf. sec. 4.3). From the mammal fauna of North America (Fox, 2000), a change from a low seasonal Early Miocene climate towards a more highly seasonal climate during the Late Miocene is suggested. Fox (2000) demonstrate an increase in aridity in North America during the Late Miocene with a distinct wet season. This pattern is not observed from the PalVeg Tortonian run. For the east coast of North America during the Late Miocene, the climate changes from warm temperate to cool temperate conditions (McCARTAN ET AL., 1990), which is qualitatively consistent to the PalVeg Tortonian run. Since the end of the Middle Miocene, summer temperatures of North America decrease (WOLFE, 1994a). Coastal regions of western North America at around 13 Ma indicate some higher MATs than the interior of the North American continent (WOLFE, 1994a), which is consistent to the PalVeg Tortonian run.

In the high latitudes, generally warmer climatic conditions are observed in the PalVeg Tortonian run as compared to the Recent Control run (cf. sec. 4.3). However, these temperatures are still lower than proxy data suggest (cf. sec. 5.2). From 12 to 8 Ma, temperatures decline in Alaska (WOLFE, 1994b). During this period, the decreasing summer temperatures in the Alaska region are more affected than the winter temperatures (WOLFE, 1994b). At about 6 to 5 Ma, the Alaska summer temperatures are almost the same as today (WOLFE, 1994b). For Canada and Alaska between 9.7 and 5.7 Ma, WHITE ET AL. (1997) suggest a cooling trend. In Beringia, the Late Miocene mean annual temperatures are +2 °C to +4 °C higher than nowaydays (WOLFE, 1994b). This temperature difference is consistent to the difference between the PalVeg Tortonian and the Recent Control run (cf. sec. 4.3).

Asia

As previously mentioned (cf. sec. 4.2), the Asian summer monsoon is strengthened in the PalVeg Tortonian run as compared to the Standard Tortonian run. Due to the effects of vegetation and orography, the PalVeg Tortonian run represents warmer and more arid conditions than today in southern parts of Asia (cf. sec. 4.3). At 8 to 7 Ma, a drying of the Quaidam Basin and northern Pakistan is related to a strengthening of the monsoon system (WANG ET AL., 1999). The Asian monsoon is stated to begin at 8 to 7 Ma (GRIFFIN, 2002). At 10 Ma, the Asian summer monsoon is mentioned to exist (SAKAI, 1997) and possibly stronger than during the Holocene (DING ET AL., 1999). During the Latest Miocene, the winter monsoon is not noticeable (WU ET AL., 1998). Thus, the PalVeg Tortonian run demonstrates a good representation of the Asian monsoon during the Miocene.

South America

For the South American continent no quantitative proxy data are available (cf. sec. 5.2). Generally, the PalVeg Tortonian run suggests slightly warmer and more humid conditions than today (cf. sec. 4.3). In the Bolivian Andes at 10 Ma, evidence of a widespread grassland vegetation with temperate or even tropical grasses is derived from $\delta^{13}C$ values of fossil mammal teeth (MACFADDEN ET AL., 1994). In the Central Andes at the same time, the terrigenous influx to the Amazon fan increases (GREGORY-WODZICKI, 2000). This indicates a higher rainfall in this region as also represented in the PalVeg Tortonian run (cf. sec. 4.3).

For regions of South America south of 30°S, the PalVeg Tortonian run demonstrates more arid conditions than today (cf. sec. 4.3). For the Atacama desert between 14.7 and 8.7 Ma, an onset of aridity with a transition to hyperarid conditions is mentioned (GREGORY-WODZICKI, 2000).

Australia

Australia is almost as warm as today in the PalVeg Tortonian simulation (cf. sec. 4.3.1). The PalVeg Tortonian run demonstrates higher rainfall rates than today (cf. sec. 4.3.4), but it still indicates too arid conditions as compared to southern Australian proxy data (cf. sec. 5.2). From Late Miocene carbonate deposits at the north-east of Australia, non-tropical conditions are suggested (BETZLER, 1997). For north-east Australia, surface water temperatures of about 17 °C to 19 °C are estimated as coral reef growth is absent (BETZLER ET AL., 1995). These are cooler conditions than in the PalVeg Tortonian run.

The Northern Hemisphere's ice cover

On the Northern Hemisphere, the PalVeg Tortonian run demonstrates the lowest amount of sea ice as compared to the Standard Tortonian and Recent Control run (cf. sec. 4.2.1). For the Miocene, the Arctic sea ice volume tends to be less than today (WOLF & THIEDE, 1991). However, detailed information about the Miocene Arctic ice cover does not exist (SCHAEFFER & SPIEGLER, 1986; THIEDE ET AL., 1998). At 11 Ma, the North Atlantic and Greenland glaciation begins (HELLAND & HOLMES, 1997). During the Late Pliocene (3.5 to 2.4 Ma), a successive onset of a large-scale glaciation of the Atlantic region occurs (KLEIVEN ET AL., 2002).

5.4 Discussion and summary

The poor Miocene data base (MOSBRUGGER & SCHILLING, 1992) is a major problem to validate model results. For most parts of the world, quantitative Tortonian proxy data are insufficiently available. Model results can be verified only for selected regions. It is questionable if single locations are really representative for a whole model grid cell. Due to the limited resolution of climate models, small scale effects are parameterised. Parameterisations such as the cumulus convection scheme of the ECHAM model (ROECKNER ET AL., 1996) include uncertainties.

The Tortonian (11 to 7 Ma) corresponds to a time span of about 4Ma. Regarding the palaeoclimate information, studies can differ as they refer to specific time intervals within the Tortonian (VAN DAM & WELTJE, 1999). On the one hand, MUDIE & HELGASON's (1983) eastern Iceland location, which is dated between 10.3 and 9.5 Ma, represents the same climatic conditions as the PalVeg Tortonian run (cf. fig. 5.1). On the other hand, a more western Iceland location is dated between 11 to 5.3 Ma (MAI, 1995) but indicates different climatic conditions than the PalVeg Tortonian run. Regarding European proxy data (VAN DAM & WELTJE, 1999), the PalVeg Tortonian run tends to correspond to the situation during the Late Tortonian.

Generally, proxy-based methods for climate reconstructions such as the used CLAMP (WOLFE, 1993) or coexistence approach (MOSBRUGGER & UTESCHER, 1997) include uncertainties. The CLAMP method is calibrated on climate observations and leaf distributions predominantly from North America (WOLFE, 1993). It is questionable, if this method can be used for palaeoclimate reconstructions of European regions. When applied to Late Oligocene to Middle Miocene European locations, TRAISER (submitted) demonstrates that the CLAMP method produces systematically lower mean annual temperatures than the coexistence approach. Despite such shortcomings, CLAMP achieves reasonable results (GREENWOOD & WING, 1995; WOLFE, 1993; WOLFE, 1994a).

The coexistence approach (MOSBRUGGER & UTESCHER, 1997) also includes uncertainties. During the Cretaceous or Eocene, the tropics are assumed to be warmer than present (PEARSON ET AL., 2001). Modern plant communities can only represent Recent climatic conditions. From the relationship between modern and fossil plants, it is not possible to estimate the upper temperature limit of such 'supertropic' palaeosituations. Thus, the coexistence approach cannot be applied to situations whose conditions are not found today. But as the Late Miocene climate is basically comparable to the modern situation (BRUCH, 1998), this argument does not apply for the present study. However, it can be difficult to find a next living relative. Consequently, the climate interval of a particular taxon can be (more or less) incorrect. The coexistence approach uses the joint intersection interval of all taxa, which keeps single errors small. Similar to the CLAMP method, the coexistence approach demonstrates its capability for palaeoclimate reconstructions (BRUCH, 1998; BRUCH & MOSBRUGGER, 2002; UTESCHER ET AL., 1997; UTESCHER ET AL., 2000).

When compared to proxy data, the PalVeg Tortonian run demonstrates more realistic results than the Standard Tortonian run. The PalVeg Tortonian run still reveals some shortcomings such as a too steep meridional temperature gradient, which is reflected in too cool high latitudes (cf. fig. 5.2). For Beringia, a temperature difference of +2 °C to +4 °C is presented in the PalVeg Tortonian run as compared to the Recent Control run (cf. sec. 4.3.1). During the Late Miocene, proxy data (WOLFE, 1994b) also suggest Beringian temperatures being +2 °C to +4 °C higher than today. Considering the precipitation patterns of the tropics, the difference is assumed to be reasonable between the PalVeg Tortonian run and the Recent Control run (cf. sec. 4.3.4). The PalVeg Tortonian run demonstrates an unrealistically high tropical rainfall when compared to proxy data (cf. sec. 5.2). The direct comparison of the PalVeg Tortonian run with proxy data is problematic. But tendencies, the PalVeg Tortonian minus Recent Control run and Tortonian proxy data minus modern observation data, are quite consistent.

As compared to the Standard Tortonian run, the PalVeg Tortonian simulation indicates higher precipitation rates in the high latitudes (cf. sec. 4.2.3). As a higher freshwater influx in the ocean produces lower $\delta^{18}O$ values, the palaeo-sea surface temperatures, which are used to determine the ocean heat transport (STEPPUHN ET AL., 2006), can be overestimated in the high latitudes. Other model studies support that SSTs are overestimated, if the freshwater influx increases (WERNER ET AL., 1999). Contrarily, the palaeo-SSTs can be underestimated in the low latitudes, as the $\delta^{18}O$ rises because of an increased evaporation in the PalVeg Tortonian run as compared to the Standard Tortonian run. For the Late Cretaceous and the Eocene, underestimated tropical SSTs are mentioned from PEARSON ET AL. (2001). Tropical SSTs are almost as high (or even higher) as today during the Late Cretaceous and the Eocene (PEARSON ET AL., 2001). This indicates that, if STEPPUHN ET AL. (2006) underestimate tropical SSTs and overestimate the SSTs in high latitudes, the reconstructed Tortonian meridional temperature gradient is too flat. Hence, STEPPUHN ET AL. (2006) underestimate the oceanic heat transport. This can explain the too cool high latitudes in the PalVeg Tortonian run (cf. sec. 5.2).

The reconstructed oceanic heat transport has to be scrutinised, as the reconstructed palaeo-SSTs are probably unrealistic. Particularly the North Pacific region tends to be too cool in the PalVeg Tortonian run, whereas the North Atlantic Ocean is represented quite well (cf. sec. 5.2). Thus, the heat transport in the Atlantic Ocean can be realistic, but not in the Pacific Ocean. Considering the data base for the reconstruction of the oceanic heat transport (STEPPUHN ET AL., 2006), the SSTs of the high latitudes are almost exclusively based on isotope data from the North Atlantic. Due to this lack of isotope data, the palaeoceanic heat transport in the Pacific Ocean could be underestimated. A strengthened heat transport in the Pacific Ocean, as compared to the prescribed weakened one, could produce some warmer conditions in the high latitudes of the Pacific region such as in Siberia. As a result, discrepancies between model and proxy data should become smaller.

Regarding the palaeogeography, the Paratethys is not included neither in the Standard Tortonian run nor in the PalVeg Tortonian run (cf. sec. 2.2.1 and 4.1). RAMSTEIN ET AL. (1997) demonstrate that the shrinking of the Paratethys during the Cenozoic contributes to a cooling of the Central Eurasia climate and particularly in Siberia (RAMSTEIN ET AL., 1997). Due to the Paratethys, the former times Eastern European climate is more humid than today (RAMSTEIN ET AL., 1997). Thus, inconsistencies particularly in Siberia and the Mediterranean region can be attributed to a missing Paratethys in the PalVeg Tortonian run.

6 VEGETATION MODELLING WITH CARAIB

With a proxy-based reconstruction of the Tortonian vegetation (cf. sec. 3), an ECHAM4/ML simulation, the PalVeg Tortonian run, was performed (cf. sec. 4). On the one hand, the vegetation-induced effects on climate were figured out when the PalVeg Tortonian run was compared with the Standard Tortonian run (cf. sec. 4.2). On the other hand, differences between the Tortonian and the present climate were pointed out (cf. sec. 4.3). The model results of the PalVeg Tortonian run were validated with proxy data (cf. sec. 5). In the present section, the climatological fields of the PalVeg Tortonian run are used to run the carbon cycle and vegetation model CARAIB (FRANCOIS ET AL., 2006). CARAIB simulates the Tortonian vegetation, which is compared to the proxy-based reconstruction of the palaeovegetation. In the case of discrepancies, it has to be considered whether models are not sufficiently able to simulate palaeoclimate (cf. sec. 5) and/or palaeovegetation or the proxy-based reconstruction has to be revised. With a lower-/higher-than-present atmospheric CO_2 in the CARAIB simulations, the fertilisation effect on the Tortonian vegetation is investigated. Finally, CARAIB allows to simulate differences between the Tortonian and the present-day's carbon cycle.

6.1 The CARAIB model and its setup for the Tortonian

In its original version, CARAIB is a pure carbon cycle model, which describes processes like stomatal regulation, C_3 and C_4 photosynthesis or organic matter decomposition in the soil (GERARD ET AL., 1999; NEMRY ET AL., 1996; WARNANT ET AL., 1994). Later on, plant functional types (PFTs) are added as a further module (OTTO ET AL., 2002). PFTs calculated from the net primary production (NPP) are used to obtain biomes. CARAIB is a combined carbon cycle and vegetation model, which is able to reproduce the Recent vegetation patterns quite well (OTTO ET AL., 2002). It is applied to past periods such as the last glacial maximum (OTTO ET AL., 2002) or the Eocene (FRANCOIS, pers. comm.). Within the EEDEN (Environments and Ecosystem Dynamics of the Eurasian Neogene) framework funded by the ESF (European Science Foundation), the CARAIB model is applied to the Tortonian using data of the PalVeg Tortonian run (cf. sec. 4). Regarding the Late Miocene carbon cycle and model-based reconstruction of the Tortonian vegetation, the main aspects of the CARAIB modelling are summarised (FRANCOIS ET AL., 2006).

In order to apply CARAIB to the Tortonian, climatological 10-year-average data from the PalVeg Tortonian run and the Recent Control experiment with ECHAM4/ML are provided (cf. sec. 4). These data include monthly means of the near-surface 2m-temperature, precipitation, cloud cover (inversely giving the sunshine hours), air relative humidity and near-surface wind speed in the standard resolution of T30 (3.75° × 3.75°). However, CARAIB does not use the real model output, but anomalies to climatological reference fields (FRANCOIS ET AL., 1999). The differences between the PalVeg Tortonian and the Control run are used, not the absolute values. To obtain the CARAIB input fields, a higher resolved present-day's observation dataset (2.5° × 2.5°) is transformed to the corresponding ECHAM T30 grid (FRANCOIS ET AL., 2006). The deviations of the PalVeg Tortonian run with respect to the Control run (Δq) are added to the reference data base (q_{obs}) yielding the CARAIB input fields (q):

$$q = q_{obs} + \Delta q.$$

The atmospheric CO_2 has to be prescribed in the CARAIB model. For the PalVeg Tortonian run, the atmospheric CO_2 is set to the Recent level of 353 ppm (cf. sec. 4.1). Without altering the climate forcing, different CO_2 concentrations can be considered in the CARAIB model. The biogeography module of CARAIB is calibrated with the pre-industrial CO_2 of 280 ppm (OTTO ET AL., 2002). In accordance to the CARAIB control simulation *PD280*, the pCO_2 of the first Tortonian CARAIB run (referred to as *Torton280*) is set to 280 ppm. To figure out the fertilisation effect of CO_2, a low CO_2 (200 ppm) and a high CO_2 (560 ppm) sensitivity experiment with CARAIB are performed (FRANCOIS ET AL., 2006). These runs are referred to as *Torton200* and *Torton560*, respectively.

6.2 Results of CARAIB simulations

6.2.1 The simulated vegetation

In contrast to the modern Greenland glaciers of PD280, Torton280 demonstrates interior grasslands and boreal forest along coastal areas (fig. 6.1). This is in accordance with the warmer and more humid conditions in Greenland in the PalVeg Tortonian run as compared to the Recent Control run (cf. sec. 4.3). Primarily due to the lower elevation of Greenland during the Late Miocene (cf. sec. 2.1.1), the temperature increase is rather strong in the PalVeg Tortonian run as compared to today (cf. sec. 4.3.1). Increases in precipitation are just moderate in the PalVeg Tortonian run (cf. sec. 4.3.4), which prevents the occurrence of boreal forests in

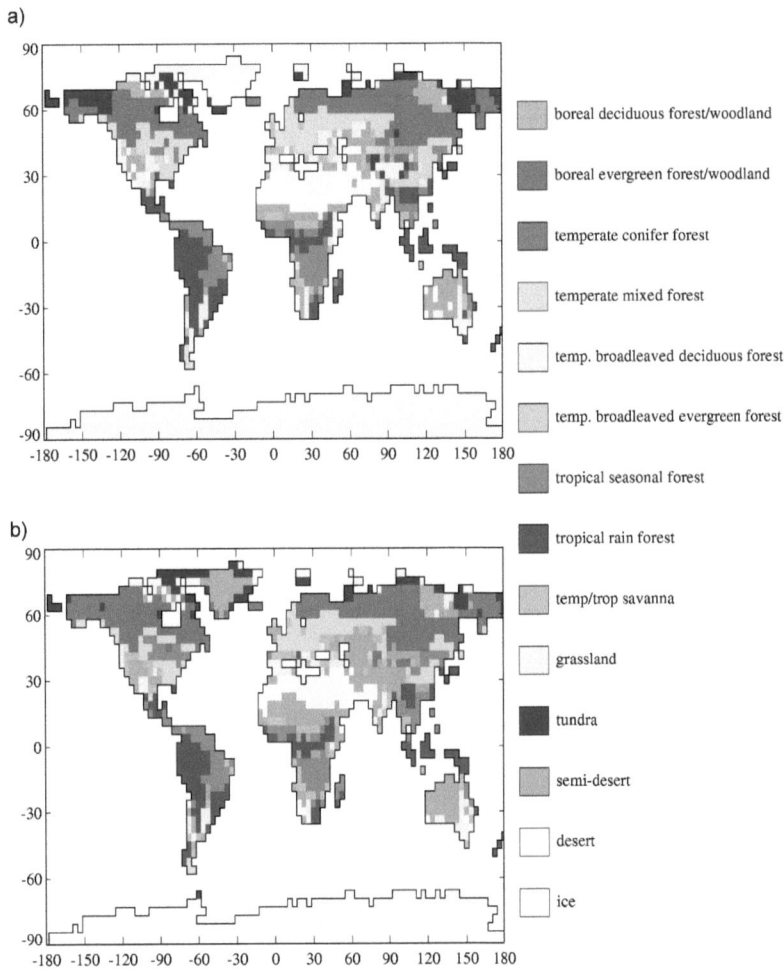

Figure 6.1: *The biome distribution of a) the PD280 and b) the Torton280 simulation with CARAIB (modified from FRANCOIS ET AL., 2006).*

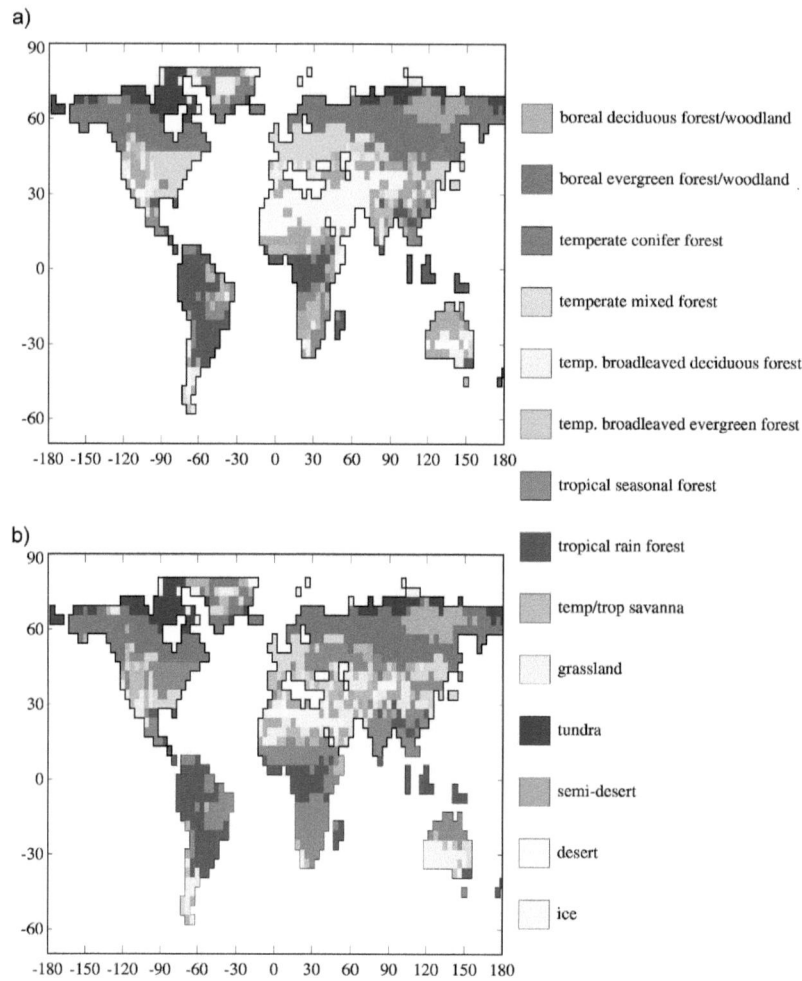

Figure 6.2: *The biome distribution of a) the Torton200 (low CO_2) and b) the Torton560 (high CO_2) simulation with CARAIB (modified from FRANCOIS ET AL., 2006).*

the interior of Greenland in Torton280. For the Himalayan, which has about half of its Recent height during the Tortonian (sec. 2.1.1), mostly grassland but also some semi-desert and boreal forest cover exist in Torton280 instead of PD280's ice and tundra vegetation. Globally, there is less evidence of extreme deserts in Torton280. Tropical seasonal and warm temperate forests are more relevant in Torton280 as compared to PD280 (fig. 6.1). This pattern is quite similar to the proxy-based reconstructed Tortonian vegetation (cf. fig. 3.2a). Due to the higher rainfall rates of the PalVeg Tortonian run (cf. sec. 4.2.3), semi-desert and tropical seasonal forest areas grow at the expense of deserts in Torton280 as compared to PD280 (fig. 6.1).

Fig. 6.3 quantitatively illustrates the change in vegetation from the Tortonian (Torton280) till today (PD280). PD280's deserts are almost halved in Torton280 (-11.4×10^6 km^2) and are replaced by a larger extension of tropical seasonal forests ($+10.3 \times 10^6$ km^2) than in PD280. Ice cover and tundra are reduced by -2.4×10^6 km^2 and -2.6×10^6 km^2, respectively, in Torton280 as compared to PD280 (fig. 6.3). In turn, semi-desert areas ($+2.8 \times 10^6$ km^2) and boreal evergreen forests ($+3.8 \times 10^6$ km^2) are more extended in Torton280.

Regarding forest, grassland and desert type biomes, the CARAIB vegetation of Torton280 and PD280 (fig. 6.3) can be compared with the proxy-based reconstruction of the Tortonian vegetation and the modern one as obtained from the Prentice biome model (cf. fig3.3). For both cases, the trend between the Tortonian and the modern vegetation is consistent. Forests in Torton280 cover 65 % of the Earth's surface. In the proxy-based reconstruction, the forest cover is 82 %, a difference of +17 % to Torton280. This is consistent with the too cool conditions in the PalVeg Tortonian run (cf. sec. 5). The present-day's forest cover of PD280 and the "Prentice distribution" is quite similar (about 55 %). Torton280 demonstrates less grassland (–2.4 %) than PD280 (fig. 6.3). Contrarily, the proxy-based Tortonian reconstruction represents larger extended graslands (+8 %) than today (cf. fig. 3.3). The desert cover of Torton280 (18 %) is larger than the one of the proxy-based reconstruction (1 %).

From the lower CO_2 (200 ppm) scenario Torton200 and the higher CO_2 (560 ppm) scenario Torton560, the fertilisation effect of atmospheric CO_2 can be estimated (fig. 6.2). In Torton560 (fig. 6.2b), the amount of deserts globally decreases as compared to PD280 and Torton280 (fig. 6.1). Grassland or savanna biomes expand at the expense of deserts in Torton560 as compared to Torton280. When increasing atmospheric CO_2, tropical seasonal forests are more important in tropical and temperate climate zones. Contrarily, Torton200 (fig. 6.2a) demonstrates the lowest amount of tropical seasonal forest of all Tortonian CARAIB simulations and also as compared to PD280 (fig. 6.1a). Tropical seasonal forests are reduced by -13.3×10^6 km^2 in Torton200 when compared to Torton280. Between Torton560 and Torton280, seasonal forests increase by $+12.8 \times 10^6$ km^2 (fig. 6.3). Accordingly, the areas of deserts and semi-deserts are larger ($+10.0 \times 10^6$ km^2 and $+5.9 \times 10^6$ km^2) at low CO_2 as compared to Torton280. Torton560 as compared to Torton280 demonstrates a reduced desert and semi-desert cover (-9.7×10^6 km^2 and -4.3×10^6 km^2). The more frequent occurrence of tropical seasonal forests in Torton560 with respect to Torton280 can be explained by the dependence of the net primary productivity on the availability of water. This dependence is reduced when stomata close at higher CO_2 concentrations. In the case of low CO_2 (Torton200), this mechanism is inverted. Thus, tropical seasonal forests have an advantage/disadvantage as compared to other biomes in water stress regions if CO_2 is high/low.

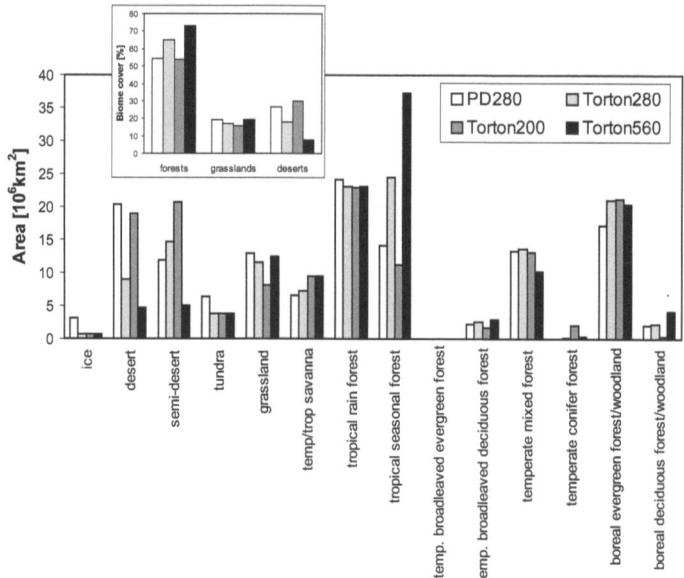

Figure 6.3: *The global biome area [10^6 km^2] of the CARAIB simulations PD280, Torton280, Torton200 and Torton560 (modified from FRANCOIS ET AL., 2006). The biome cover [%] for the summed forest, grassland and desert type biomes is additionally represented in the smaller figure (cf. fig. 3.3).*

6.2.2 The carbon cycle

The vegetation is affected by the different climate in the PalVeg Tortonian run compared to the Recent Control simulation. Hence, the carbon storage is affected (tab. 6.1). In total, the carbon stock of Torton280 is modestly higher (+159 Gt C) than for PD280 with nearly the same contributions of the biosphere (+84 Gt C) and the soil (+75 Gt C) storage. The carbon stock of tropical seasonal forests in Torton280 rises (+188 Gt C) as compared to PD280 (fig. 6.4). Within boreal evergreen forests, the storage increases by about +77 Gt C in Torton280. The carbon stock decreases for the biomes tundra (–58 Gt C), tropical rain forest (–36 Gt C) and grassland (–20 Gt C) in Torton280 in relation to PD280. For other biome types, changes in the total carbon stock of Torton280 as compared to PD280 are of minor importance (fig. 6.4).

From tab. 6.1, the carbon storage is clearly shown to depend on atmospheric CO_2. As compared to Torton280, the total carbon stock of Torton200 is reduced (–831 Gt C), while it increases

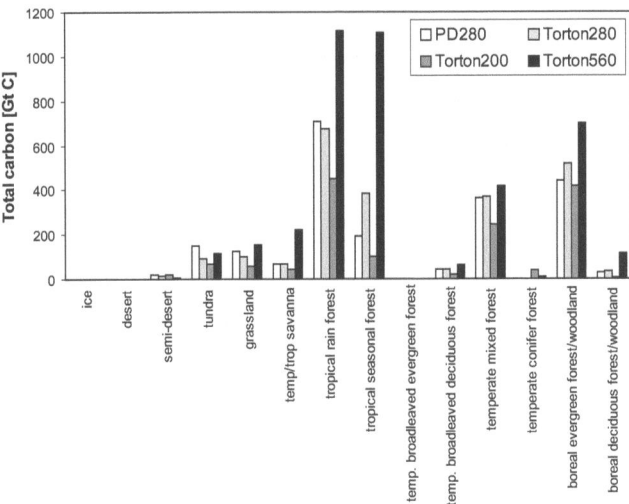

Figure 6.4: *The total carbon storage [Gt C] of the CARAIB simulations PD280, Torton280, Torton200 and Torton560 (modified from FRANCOIS ET AL., 2006).*

in Torton560 (+1727 Gt C). The importance of the atmospheric CO_2 for palaeovegetation reconstructions is evident, even though the fertilisation effect can be overestimated because of the absence of nutrient cycles in the CARAIB model. The carbon stocks (fig. 6.4) of tropical and tropical seasonal forests increase (+441 Gt C and +723 Gt C) in Torton560 as compared to Torton280. For these biomes, the carbon stocks are reduced (–226 Gt C and –280 Gt C) in Torton200. For other biome types, a sensitivity of the carbon stock to pCO_2 can also be observed. For example, the carbon storage is slightly higher in cases of boreal evergreen forest and in savanna (+183 Gt C and +152 Gt C) in Torton560 as compared to Torton280. For the same biomes, the carbon stock decreases (–100 Gt C and –25 Gt C) in Torton200 as compared to Torton280. From differences between Torton280 and PD280 and between Torton560 and Torton200, it can be concluded that variations in CO_2 are more important for the carbon cycle than differences between the Tortonian and present-day's climate.

6.3 Discussion

Basically, FRANCOIS ET AL.'s (2006) CARAIB simulation Torton280 (cf. sec. 6.2.1) supports some main characteristic of the proxy-based reconstruction of the Tortonian vegetation (cf. sec. 3). The extension of boreal forests far into the Arctic high latitudes during the Tortonian is also in accordance with other proxy data (BOULTER & MANUM, 1997). The model-based palaeovegetation reconstructions demonstrate an expansion of tropical forests (particularly

	vegetation	soil	total
PD280	900	1228	2128
Torton280	984	1303	2287
Torton280 – PD280	+84	+75	+159
Torton200 (low CO_2)	626	830	1456
Torton560 (high CO_2)	1721	2293	4014
Torton560 – Torton200	+1095	+1463	+2558

Table 6.1: *The global carbon stocks [Gt C] of the vegetation, the soil and in total for the CARAIB simulations PD280, Torton280, Torton200, Torton560 and the differences between Torton280 and PD280 and between Torton560 and Torton200, respectively (modified from FRANCOIS ET AL., 2006).*

tropical seasonal forests in Torton280) at the expense of deserts. This is consistent with the proxy-based reconstruction (cf. sec. 3) as well as other proxy data (HOORN ET AL., 2000; IVANOV ET AL., 2002; STRÖMBERG, 2002).

There are differences between Torton280 and the proxy-based reconstruction (cf. sec. 6.2.1). It should be noticed that a different vegetation classification is used in both cases. The used Prentice biomes (PRENTICE ET AL., 1992) differ from those of CARAIB (FRANCOIS ET AL., 2006). For instance, the area of the Recent Sahara desert is characterised as a warm grassland in the proxy-based Tortonian reconstruction (cf. fig. 3.2a), whereas it is partly semi-desert in Torton280. It is the question, whether a biome 'semi-desert' is comparable to a biome 'warm grass' or not.

The CARAIB model demonstrates a strong sensitivity to the carbon dioxide forcing (cf. sec. 6.2). Some differences between Torton280 and the proxy-based reconstruction can be caused by an underestimated CO_2 in the CARAIB simulation (280 ppm). However, there is more evidence for an atmospheric CO_2 at approximately the pre-industrial level or even lower during the Late Miocene (PEARSON & PALMER, 2000).

Regarding the carbon cycle modelling (cf. sec. 6.2.2), the soil carbon stock of PD280 agrees with other Recent studies (ADAMS ET AL., 1990; FRANCOIS ET AL., 1998; FRANCOIS ET AL., 1999; PRENTICE & FUNG, 1990). PD280 demonstrates a vegetation carbon stock (900 Gt C), which is within the upper range of other author's estimations (ADAMS ET AL., 1990; FRANCOIS ET AL., 1998; FRANCOIS ET AL., 1999; PRENTICE & FUNG, 1990). Additionally, PD280 represents the natural equilibrium. Studies, which suggest a today's value of 610 Gt C for the vegetation carbon, include the human influence (CLIMATE CHANGE, 1995). FRANCOIS ET AL. (2006) conclude that an overestimation of tropical forests in the CARAIB model produces some differences in the CARAIB carbon stock as compared to other studies (HAXELTINE & PRENTICE, 1996; MATTHEWS, 1983; MELLILO ET AL., 1993). The natural biomass burning is a regulation process within the carbon cycle (THONICKE ET AL, 2001), which is missing in CARAIB and causes some discrepancies. A present-day's CARAIB simulation in the higher resolution of $0.5° \times 0.5°$ demonstrates a vegetation carbon stock of 761 Gt C (OTTO ET AL., 2002). Thus, the performance of the CARAIB model also depends on the used spatial resolution (FRANCOIS ET AL., 2006).

Regarding the variation of the total carbon stock between Torton280 and PD280 (+159 Gt C), this value appears to be quite small (cf. Sec. 6.2.2). For the last glacial maximum (LGM), CARAIB demonstrates a difference in the total carbon storage of –840 Gt C in the LGM simulation as compared to the modern control experiment (OTTO ET AL., 2002). This variation is attributed to the climate change from the LGM till today (OTTO ET AL., 2002). A range of the total carbon stock is spanned from 1891 Gt C for a pre-industrial scenario (PRENTICE & FUNG, 1990) to present-day's 2190 Gt C (CLIMATE CHANGE, 1995). This range of +299 Gt C is almost twice as high as the difference between Torton280 and PD280 (+159 Gt C). However, the absolute total carbon stock of Torton280 (2287 Gt C) is higher than for the above mentioned Recent studies (CLIMATE CHANGE, 1995; PRENTICE & FUNG, 1990).

7 SUMMARY AND CONCLUSIONS

For the present PhD thesis, a proxy-based reconstruction of the Tortonian vegetation (cf. sec. 3) is used to perform a more realistic climate model simulation for the Tortonian, the so-called PalVeg Tortonian run (cf. sec. 4), using the ECHAM4/ML model. The Tortonian vegetation (cf. fig. 3.2) is characterised by a larger amount of forests as compared to the modern vegetation. During the Tortonian, forests extend far towards the high latitudes of the Northern Hemisphere. Today's widespread desert areas such as the Sahara desert of North Africa almost vanish in the Tortonian. As a consequence, the global albedo of the PalVeg Tortonian run decreases (–40 %) as compared to the Standard Tortonian run. The global average leaf area index and the maximum available soil water capacity increases by +83 % and +8 %, respectively, in the PalVeg Tortonian simulation (cf. sec. 4.2.1).

The effects of the Tortonian vegetation

Due to the adjustment of surface parameters applying to vegetation, the PalVeg Tortonian run demonstrates a strong influence of the Tortonian vegetation on the climate system as compared to the Standard Tortonian run (cf. sec. 4). The PalVeg Tortonian run indicates globally warmer (+0.9 °C) and more humid (+36 mm/a) conditions than the Standard Tortonian run (cf. sec. 4.2.1). Particularly the high northern latitudes are warmer in the PalVeg Tortonian run than in the Standard Tortonian run. Accordingly, the PalVeg Tortonian run demonstrates the shallowest meridional temperature gradient (cf. sec. 4.2.2) as well as the lowest Arctic sea ice volume (cf. sec. 4.2.1) when compared to the Recent Control run and the Standard Tortonian run.

The warming in the PalVeg Tortonian run occurs primarily over continental areas (cf. sec. 4.2.3). In the today's cold-continental high latitudes, PalVeg Tortonian run demonstrates the maximum increases in the mean annual temperatures (+4 °C) run, which is attributed to the replacement of the Recent tundra vegetation with boreal forests as compared to the Standard Tortonian. The winter temperatures in the high latitudes raise with maximally +7 °C in the PalVeg Tortonian run, while summer temperatures just modestly increase as compared to the Standard Tortonian run. In the high latitudes, the seasonal temperature cycle is reduced in the PalVeg Tortonian as compared to the Standard Tortonian run. For North Africa, the summer temperatures are lower in the PalVeg Tortonian than in the Standard Tortonian simulation.

This summer cooling is attributed to the replacement of the Recent Sahara desert with warm grassland.

The water cycle globally intensifies in the PalVeg Tortonian run. Over the continents, variations in precipitation and evapotranspiration are more pronounced than over the oceans in the PalVeg Tortonian run (cf. sec. 4.2.3). The continents demonstrate increased evapotranspiration and precipitation rates as compared to the Standard Tortonian reference simulation. In North Africa and Asia, the annual precipitation is higher by more than +400 mm/a in the PalVeg Tortonian simulation. Regarding the absolute precipitation rates in the high latitudes, the relative increase rates are remarkable in the PalVeg Tortonian run. Beyond the influence of the palaeovegetation on continental areas, a moderate warming occurs over ocean surfaces in the PalVeg Tortonian run (cf. sec. 4.2.3). The PalVeg Tortonian run demonstrates an increased moisture transport from equatorial regions towards higher latitudes over the oceans and from the oceans to the land surfaces.

As a consequence of the larger energy supply from the surface in the PalVeg Tortonian run, the atmospheric circulation regimes are affected (cf. sec. 4.2.4). Particularly a strengthened northward heat transport in the atmosphere contributes to a reduction of the equator-to-pole temperature difference in the PalVeg Tortonian run as compared to the Standard Tortonian run. On the regional scale, the change of the Recent to the palaeovegetation causes a strengthened Asian summer monsoon in the PalVeg Tortonian run (cf. sec. 4.2.5). The Tortonian Asian monsoon is, however, still weaker than nowadays, which is in accordance to proxy data (SAKAI, 1997; WU ET AL., 1998). For North Africa, the PalVeg Tortonian run demonstrates a northward shift of the Hadley cell particularly during summer as compared to the Standard Tortonian run. The PalVeg Tortonian run indicates a strengthened African summer monsoon as compared to the Standard Tortonian run. For Europe and the North Atlantic Ocean, the PalVeg Tortonian run represents a high NAO index situation as compared to the Standard Tortonian run. The North Atlantic storm track regimes displace towards the north in the PalVeg Tortonian simulation. This causes an increased temperature and moisture advection towards Northern Europe. The shift of the storm tracks contributes to a more efficient northward heat transport in the PalVeg Tortonian run. This reduces the meridional temperature gradient in the PalVeg Tortonian run.

Validation of the PalVeg Tortonian run with terrestrial proxy data

In comparison to terrestrial proxy data, the PalVeg Tortonian run demonstrates a generally better performance than the Standard Tortonian run (cf. sec. 5). The precipitation pattern of the PalVeg Tortonian run agrees with proxy data. But in the tropics, the annual rainfalls are overestimated in the PalVeg Tortonian run. This weak point is attributed to the ECHAM model itself and not to the Tortonian boundary conditions. The simulated mean annual temperatures are more realistic in the PalVeg Tortonian run than those in the Standard Tortonian run. In the high latitudes and in Central Europe, the PalVeg Tortonian run indicates cooler conditions than proxy data suggest. For the North Pacific region, the too cool conditions in PalVeg Tortonian run demonstrate that the palaeoceanic heat transport could be underestimated. The meridional temperature gradient is too steep in the PalVeg Tortonian run as compared to proxy data.

CARAIB modelling

The CARAIB model (cf. sec. 6) reproduces the main features of the proxy-based reconstruction of the Tortonian vegetation (cf. sec. 3). This demonstrates that the models ECHAM4/ML and CARAIB simulate the palaeoclimate and palaeovegetation quite realistic. Some insufficiencies in the simulated Tortonian vegetation represent weak points in the PalVeg Tortonian run. Other weak points are due to the CARAIB model. Regarding North Africa, the CARAIB model represents rather more desert and semi-desert conditions than warm grassland, as it is indicated from the proxy-based reconstruction. CARAIB sensitivity experiments demonstrate a strong sensitivity to the CO_2 forcing: Variations in CO_2 are even more important for the vegetation than the climatic differences between the PalVeg Tortonian run and the Recent Control experiment.

Conclusions

In general, ECHAM4/ML simulates a quite realistic Tortonian climate, when considering the effects of a weaker ocean heat transport and the lower palaeorography as well as the palaeovegetation (tab. 7.1). However, some insufficiencies such as a meridional temperature gradient, which is steeper than suggested from proxy data, still remain in the PalVeg Tortonian run. These shortcomings can partly be attributed to effects of an unrealistic palaeogeography in the model setup. For instance, the Paratethys is not included neither in the Standard Tortonian run (cf. sec. 2.2; STEPPUHN ET AL., 2006) nor in the PalVeg Tortonian run (cf. sec. 4.1). RAMSTEIN ET AL. (1997) emphasise the contribution of the Paratethys in warming northern parts of Asia and Eastern Eurasia being more humid. A quite crucial parameter is also the ocean heat transport. The PalVeg Tortonian run indicates that the northward heat transport in the Pacific Ocean could be underestimated (cf. sec. 5.3). The method to adapt the flux correction of the mixed-layer ocean model includes some simplifications (STEPPUHN ET AL., 2006), which can lead to an underestimated palaeoceanic heat transport in the Pacific region. During the Late Miocene, the atmospheric CO_2 is comparable to the pre-industrial level (280 ppm) or even lower (PEARSON & PALMER, 2000). For the Tortonian simulations with ECHAM4/ML, a higher CO_2 (353 ppm) is assumed. If the assumed Tortonian CO_2 concentration is overestimated in the PalVeg Tortonian run, discrepancies between the model and proxy data should increase.

So far, the Late Miocene climate with its shallower-than-present meridional temperature gradient is still poorly understood. For the Late Miocene, further model simulations should be performed using coupled atmosphere-ocean general circulation models. To estimate the influence of the atmosphere on the ocean circulation, ECHAM4/ML model data for the Tortonian should also be used to run an ocean circulation model. In turn, Miocene OGCM results for the oceanic heat transport should be used to perform further Tortonian AGCM simulations. If the Standard Tortonian run and the PalVeg Tortonian run are used to run an OGCM, this would allow to estimate the influence of the palaeovegetation on the ocean circulation.

	Standard Tortonian run (STEPPUHN ET AL., 2006)	PalVeg Tortonian run
global temperature	–	O
- high latitudes	– –	–
- mid-latitudes	– –	O
- low latitudes	O	O
precipitation	–	O
Arctic sea ice volume	+ +	+
meridional temperature gradient	+ +	+

Table 7.1: *The summarised qualitative agreements (O) and the over-/underestimations (+/–) of the model results of the Standard Tortonian run (cf. tab. 2.2) and the PalVeg Tortonian run as compared to proxy data.*

ACKNOWLEDGEMENTS

This work was financially supported by the Deutsche Forschungsgemeinschaft (DFG). Prof. Dr. Volker Mosbrugger and his working group, amongst others Dr. Angela A. Bruch, Dr. Anke Steppuhn, Dr. Torsten Utescher and Dr. Dieter Uhl, is sincerely thanked for contributing and supporting my work. I would like to thank Dr. Joachim Kuhlemann and other members of the Institute of Geosciences, University of Tübingen.

For reviewing this PhD thesis, I would like to thank Prof. Dr. V. Mosbrugger and Prof. Dr. C. Hemleben.

The model ECHAM4/ML was kindly provided from the Max-Planck Institute of Meteorology, Hamburg. The ECHAM4/ML computer simulations were performed at the Deutsches Klimarechenzentrum (DKRZ) in Hamburg with special support from Dr. Erich Roeckner and Dipl.-Met. Ulrich Schlese.

Dr. Louis François and Maxime Ghislain from the Laboratoire de Physique Atmosphérique et Planétaire, University of Liège, has to be kindly thanked for the CARAIB modelling studies within the EEDEN project ("Environments and Ecosystem Dynamics of the Eurasian Neogene"), which is financially funded by the European Science Foundation (ESF).

Finally, I would like to very kindly thank Franziska Großmann for her helping comments.

REFERENCES

ADAMS, J.M., FAURE, H., FAURE-DENARD, L., MCGLADE, J.M., WOODWARD, F.I., 1990. Increases in the terrestrial carbon storage from the Last Glacial Maximum to the present. *Nature*, **348**, 711-714.

ALONSO-ZARZA, A.M., CALVO, J.P., 2000. Palustrine sedimentation in an episodically subsiding basin: the Miocene of the northern Teruel Graben (Spain). *Palaeogeography, Palaeoclimatology, Palaeoecology*, **160**, 1-21.

BARRON, E.J., WASHINGTON, W.M., 1984. The role of geographic variables in explaining paleoclimates: results from Cretaceous climate model sensitivity studies. *Journal of Geophysical Research*, **89**, 1267-1279.

BARRON, E.J., 1985. Explanations of the Tertiary global cooling trend. *Palaeogeography, Palaeoclimatology, Palaeoecology*, **50**, 45-61.

BARRON, E.J., HAY, W.W., THOMPSON, S., 1989. The hydrological cycle: A major variable during Earth history. *Palaeogeography, Palaeoclimatology, Palaeoecology*, **75**, 157-174.

BARRON, E.J., PETERSON, W.H., 1991. The Cenozoic ocean circulation based on ocean general circulation model results. *Palaeogeography, Palaeoclimatology, Palaeoecology*, **83**, 1-28.

BENSON, R.H., HAYEK, L.-A., HODELL, D.A., RAKIC-EL BIED, K., 1995. Extending the climatic precession curve back into the late Miocene by signature template comparison. *Paleoceanography*, **10(2)**, 5-20.

BERGER, A.L., 1978. Long-term variations of daily insolation and Quaternary climatic changes. *Journal of Atmospheric Sciences*, **35**, 2362-2367.

BERGER, A., LOUTRE, M.F., GALEÉ, H., 1998. Sensitivity of the LLN climate model to the astronomical and CO_2 forcings over the last 200ky. *Climate Dynamics*, **14**, 615-629.

BERICHT DES SFB 275 DER UNIVERSITÄT TÜBINGEN. Klimagekoppelte Prozesse in meso- und känozoischen Geoökosystemen. Band 1, 1998-2000.

BERNER, R.A., 1994. Geocarb II: a revised model of atmospheric CO_2 over Phanerozoic time. *American Journal of Science*, **294**, 56-91.

BERTINI, A., 1994. Palynological investigations on Upper Neogene and Lower Pleistocene sections in Central and Northern Italy. *Mem. Soc. Geol. It.*, **48**, 431-443.

BETZLER, C., BRACHERT, T.C., KROON, D., 1995. Role of climate in partial drowning of the Queensland Plateau carbonate platform (northeastern Australia). *Marine Geology*, **123**, 11-32.

BETZLER, C., 1997. Ecological Controls on Geometries of Carbonate Platforms: Miocene/Pliocene Shallow-water Microfaunas and Carbonate Biofacies from the Queensland Plateau (NE Australia). *Facies*, **37**, 147-166.

BICE, K.L., SCOTESE, C.R., SEIDOV, D., BARRON, E.J., 2000. Quantifying the role of geographic change in Cenozoic ocean heat transport using uncoupled atmosphere and ocean models. *Palaeogeography, Palaeoclimatology, Palaeoecology*, **161**, 295-310.

BIGG, G.R., 1999. The oceans and climate. *Cambridge University Press*, Cambridge,

BLACKMON, M.L., 1976. A climatological spectral study of the 500mb geopotential height of the Northern Hemisphere. *Journal of the Atmospheric Sciences*, **33**, 1607-1623.

BLANC, P.-L., 2000. Of sills and straits: a quantitative assessment of the Messinian Salinity Crisis. *Deep-Sea Research*, **I 47**, 1429-1460.

BLENDER, R., FRAEDRICH, K., LUNKEIT, F., 1997. Identification of cyclone-track regimes in the North Atlantic. *Quarterly Journal of the Royal Meteorological Society*, **123**, 727-741.

BOULTER, M.C., MANUM, S.B., 1997. A lost continent in temperate Arctic. *Endevour*, **21**, 105-108.

BRACHERT, T.C., BETZLER, C., BRAGA, J.C., MARTIN, J.M., 1996. Record of climatic change in neritic carbonates: turnover in biogenic associations and depositional modes (Late Miocene, southern Spain). *Geol. Rundsch.*, **85**, 327-337.

BRUCH, A.A., 1998. Palynologische Untersuchungen im Oligozän Sloweniens – Paläo-Umwelt und Paläoklima im Ostalpenraum. *Tübinger Mikropaläontologische Mitteilungen*, **18**, 193pp.

BRUCH, A.A., MOSBRUGGER, V., 2002. Palaeoclimate versus vegetation reconstruction - palynological investigations on the Oligocene sequence of the Sava Basin, Slovenia. *Review of Palaeobotany and Palynology*, **122**, 117-141.

BRUCH, A.A., UTESCHER, T., MOSBRUGGER, V., GABRIELYAN, I., IVANOV, D.A., 2006. Late Miocene climate in the circum-Alpine realm - a quantitative analysis of terrestrial palaeofloras. *Palaeogeography, Palaeoclimatology, Palaeoecology*, **238**, 270-280.

CARILLO, A., RUTI, P.M., NAVARRA, A., 2000. Strom tracks and zonal mean flow variability: a comparison between observed and simulated data. *Climate Dynamics*, **16**, 219-228.

CERLING, T.E., 1991. Carbon dioxide in the atmosphere: evidence from Cenozoic and Mesozoic paleosols. *American Journal of Science*, **291**, 377-400.

CERLING, T.E., HARRIS, J.M., MACFADDEN, B.J., LEAKEY, M.G., QUADE, J., EISENMANNM, V., EHLERINGER, J.R., 1997. Global vegetation change through the Miocene/Pliocene boundary. *Nature*, **389**, 153-159.

CHARNEY, J.G., 1975. Dynamics of deserts and drought in the Sahel. *Quarterly Journal of the Royal Meteorological Society*, **101**, 193-202.

CLARK, P.U., PISIAS, N.G., STOCKER, T.F., WEAVER, A.J., 2002. The role of the thermohaline circulation in the abrupt climate change. *Nature*, **415**, 863-869.

CLAUSSEN, M., 1993. Shift of biome patterns due to simulated climate variability and climate change. *Max-Planck-Institut für Meteorologie*, Hamburg, Report **115**.

CLAUSSEN, M., 1994. On coupling global biome models with climate models. *Max-Planck-Institut für Meteorologie*, Hamburg, Report **131**.

CLAUSSEN, M., BROVKIN, V., GANOPOLSKI, A., KUBATZKI, C., PETOUKHOV, V., 1998. Modelling global terrestrial vegetation-climate interaction. *Phil. Trans. R. Soc. Lond. B*, **353**, 53-63.

CLAUSSEN, M., KUBATZKI, C., BROVKIN, V., GANOPOLSKI, A., 1999. Simulation of an abrupt change in Saharan vegetation in the mid-Holocene. *Geophysical Research Letters*, **26, no.14**, 2037-2040.

CLIMATE CHANGE, 1995. The science of climate change. Contribution of working group I to the second Assessment report of the Intergovernmental Panel on Climate Change.

COLLINS, L.S., COATES, A.G., BERGGREN, A.G., AUBRY, M.P., ZHANG, J., 1996. The late Miocene Panama isthmian strait. *Geology*, **24**, 687-690.

COVEY, C., THOMPSON, S.L., 1989. Testing the effects of oceanic heat transport on climate. Palaeogeography, Palaeoclimatology, *Palaeoecology*, **75**, 331-341.

CROWLEY, T.J., 1992. North Atlantic Deep Water cools the Southern Hemisphere. *Paleoceanography*, **7(4)**, 489-497.

CROWLEY, T.J., 2000. Carbon dioxide and Phanerozoic climate. In: Warm Climates in Earth History [HUBER, B.T., MACLEOD, K.G., WING, S.L. (eds.)]. *Cambridge University Press*, Cambridge, 462pp.

CROWLEY, T.J., ZACHOS, J.C., 2000. Comparison of zonal temperature profiles for past warm time periods. In: Warm Climates in Earth History [HUBER, B.T., MACLEOD, K.G., WING, S.L. (eds.)]. *Cambridge University Press*, Cambridge, 462pp.

DECONTO, R., BRADY, E.C., BERGREN, J., HAY, W.W., 2000. Late Cretaceous climate, vegetation, and ocean interactions. In: Warm Climates in Earth History [HUBER, B.T., MACLEOD, K.G., WING, S.L. (eds.)]. *Cambridge University Press*, Cambridge, 462pp.

DE NOBLET-DUCOUDRE, N., CLAUSSEN, M., PRENTICE, C., 2000. Mid-Holocene greening of the Sahara : first results of the GAIM 6000 year BP Experiment with two asynchronously coupled atmosphere/biome models. *Climate Dynamics*, **16**, 643-659.

DING, Z.L., XIONG, S.F., SUN, J.M., YANG, S.L., GU, Z.Y., LIU, T.S., 1999. Pedostratigraphy and paleomagnetism of a approximately 7.0 Ma eolian loess-red clay sequence at Lingtai, Loess Plateau, north-central China and the implications for paleomonsoon evolution. *Palaeogeography, Palaeoclimatology, Palaeoecology*, **152**, 49-66.

DKRZ MODELLBETREUUNGSGRUPPE, 1994. The ECHAM3 Atmospheric General Circulation Model. *Deutsches Klimarechenzentrum*, Hamburg, Technical Report **6**, 182pp.

DKRZ MODELLBETREUUNGSGRUPPE, 1997. ECHAM4 - Workshop Hamburg, November 25th 1996. *Deutsches Klimarechenzentrum*, Hamburg.

DOHERTY, R., KUTZBACH, J., FOLEY, J., POLLARD, D., 2000. Fully coupled climate/dynamical vegetation model simulations over Northern Africa during the mid-Holocene. *Climate Dynamics*, **16**, 561-573.

DOUVILLE, H., ROYER, J.-F., POLCHER, J., COX, P., GEDNEY, N., STEPHENSON, D.B., VALDES, P.J., 2000. Impact of CO_2 doubling on the Asian summer monsoon: Robust versus model-dependent responses. *Journal of the Meteorological Society of Japan*, **78(4)**, 421-439.

DUTTON, J.F., BARRON, E.J., 1996. Genesis sensitivity to changes in past vegetation. *Palaeoclimates*, **1**, 325-354.

DUTTON, J.F., BARRON, E.J., 1997. Miocene to present vegetation changes: A possible piece of the Cenozoic puzzle. *Geology*, **25(1)**, 39-41.

ENGVALL, I., 2003. The comparison of the water cycle of a greenhouse world to the today's situation based on results from global climate modelling. *Master thesis Applied Environmental Geosciences*, Universität Tübingen.

FLOWER, B.P., KENNETT, J.P., 1994. The middle Miocene climatic transition: East Antarctic ice sheet development, deep ocean circulation, and global carbon cycling. *Palaeogeography, Palaeoclimatology, Palaeoecology*, **108**, 537-555.

FLUTEAU, F., RAMSTEIN, G., BESSE, J., 1999. Simulating the evolution of the Asian and African monsoons during the past 30 Myr using an atmospheric general circulation model. *Journal of Geophysical Research*, **104**, 11995-12018.

Fox, D.L., 2000. Growth increments in Gomphotherium tusks and implications for the Late Miocene climate change in North America. *Palaeogeography, Palaeoclimatology, Palaeoecology*, **156**, 327-348.

FRANCOIS, L.M., DELIRE, C., WARNANT, P., MUNHOVEN, G., 1998. Modelling the glacial-interglacial changes in the continental biosphere. *Global Planetary Change*, **16-17**, 37-52.

FRANCOIS, L.M., GODDERIS, Y., WARNANT, P., RAMSTEIN, G., DE NOBLET, N., LORENZ, S., 1999. Carbon stocks and isotopic budgets of the terrestrial biosphere at mid-Holocene and last glacial maximum times. *Chem. Geol.*, **159**, 163-189.

FRANCOIS, L., GHISLAIN, M., OTTO, D., MICHEELS, A., 2006. Late Miocene vegetation reconstruction with the CARAIB model. *Palaeogeography, Palaeoclimatology, Palaeoecology*, **238**, 302-320.

GANOPOLSKI, A., KUBATZKI, C., CLAUSSEN, M., BROVKIN, V., PETOUKHOV, V., 1998a. The influence of vegetation-atmosphere-ocean interaction on climate during the Mid-Holocene. *Science*, **280**, 1916-1919.

GANOPOLSKI, A., RAHMSTORF, S., PETOUKHOV, V., CLAUSSEN, M., 1998b. Simulation of modern and glacial climates with a coupled global model of intermediate complexity. *Nature*, **391**, 351-356.

GEBKA, M., MOSBRUGGER, V., SCHILLING, H.-D., UTESCHER, T., 1999. Regional-scale palaeoclimate modelling on soft proxy-data basis - an example from the Upper Miocene of the Lower Rhine Embayment. *Palaeogeography, Palaeoclimatology, Palaeoecology*, **152**, 225-258.

GERARD, J.-C., NEMRY, B., FRANCOIS, L., WARNANT, P., 1999. The interannual change of atmospheric CO_2: contribution of subtropical ecosystems? *Geophys. Res. Lett.*, **26**, 243-246.

GRAHAM, A., 1998. Late Cretaceous and Cenozoic history of North American vegetation. *Oxford University Press*, Oxford.

GREENWOOD, D.R., WING, S.L., 1995. Eocene continental climates and latitudinal temperature gradients. *Geology*, **23(11)**, 1044-1048.

References

GREGOR H.-J., 1982. Die jungtertiären Floren Süddeutschlands, *Enke Verlag*, Stuttgart.

GREGOR, H.-J., UNGER, H.J., 1988. Bemerkungen zur Geologie und Paläontologie der Pflanzenfundstelle Aubenham bei Ampfing. *Documenta naturae*, **42**, 37-39.

GREGORY-WODZICKI, K.M., 2000. Uplift history of the Central and Northern Andes: A review. *Geological Society of America Bulletin*, **112(7)**, 1091-1105.

GRIFFIN, D.L., 2002. Aridity and humidity: two aspects of the late Miocene climate of North Africa and the Mediterranean. *Palaeogeography, Palaeoclimatology, Palaeoecology*, **182**, 65-91.

HAGEMANN, S., BOTZET, M., DÜMENIL, L., MACHENHAUER, B., 1999. Derivation of global GCM boundary conditions from 1 km land use satellite data. *Max-Planck-Institut für Meteorologie*, Hamburg, Report **289**.

HAUG, G.H., TIEDEMANN, R., 1998. Effect of the formation of the Isthmus of Panama on Atlantic Ocean thermohaline circulation. *Nature*, **393**, 673-676.

HAXELTINE, A., PRENTICE, I.C., 1996. Biome3: An equilibrium terrestrial biosphere model based on ecological constraints, resource availability, and competition among plant functional types. *Global Biogeochemical Cycles*, **10**, 693-709.

HELLAND, P.E., HOLMES, M.A., 1997. Surface textural analysis of quartz sand grains from ODP Site 918 off the southeast coast of Greenland suggests glaciation of southern Greenland at 11Ma. *Palaeogeography, Palaeoclimatology, Palaeoecology*, **135**, 109-121.

HERMAN, A.B., SPICER, R.A., 1996. Palaeobotanical evidence for a warm Cretaceous arctic ocean. *Nature*, **380**, 330-333.

HOORN, C., OHJA, T., QUADE, J., 2000. Palynological evidence for vegetation development and climate change in the Sub-Himalayan Zone (Neogene, Central Nepal). *Palaeogeography, Paleoclimatology, Palaeoecology*, **163**, 133-161.

IPCC, 2001. Climate Change 2001: The Scientific Basis. Contribution of Working Group I to the Third Assessment Report of the Intergovernmental Panel on Climate Change [HOUGHTON, J.T., Y. DING, D.J. GRIGGS, M. NOGUER, P.J. VAN DER LINDEN, X. DAI, K. MASKELL, C.A. JOHNSON (eds.)]. *Cambridge University Press*, Cambridge, 881pp.

IVANOV, D., ASHRAF, A.R., MOSBRUGGER, V., PALAMAREV, E., 2002. Palynological evidence for Miocene climate change in the Forecarpathian Basin (Central Paratethys, NW Bulgaria). Palaeogeography, Palaeoclimatology, Palaeoecology, 178, 19-37.

JACOBS, B.F., DEINO, A.L., 1996. Test of climate-leaf physiognomy regression models, their application to two Miocene floras from Kenya, and 40Ar/39Ar dating of the Late Miocene Kapturo site. *Palaeogeography, Paleoclimatology, Palaeoecology*, **123**, 259-271.

JACOBS, B.F., 1999. Estimation of rainfall variables from leaf characters in tropical Africa. Abstract St. Louis.1999. Jacobs, B.F. Estimation of rainfall variables from leaf characters. *Palaeogeography, Paleoclimatology, Palaeoecology*, **145**, 231-250.

JOUZEL, J., BARKOV, N.I., BARNOLA, J.M., 1993. Extending the Vostok ice-core record of palaeoclimate to the penultimate glacial period. *Nature*, **364**, 407-412.

KLEIDON, A., HEIMANN, M., 2000. Assessing the role of deep rooted vegetation in the climate system with model simulations: mechanisms, comparison to observations and implications for Amazonian deforestation. *Climate Dynamics*, **16**, 183-199.

KLEIVEN, H.F., JANSEN, E., FRONVAL, T., SMITH, T.M., 2002. Intensification of Northern Hemisphere glaciations in the circum Atlantic region (3.5-2.4 Ma); ice-rafted detritus evidence. *Palaeogeography, Palaeoclimatology, Palaeoecology*, **184**, 213-223.

KUBATZKI, C., MONTOYA, M., RAHMSTORF, S., GANOPOLSKI, A., CLAUSSEN, M., 2000. Comparison of the last interglacial climate simulated by a coupled global model of intermediate complexity and AOGCM. *Climate Dynamics*, **16**, 799-814.

LATIF, M., NEELIN, J.D., 1994. El Niño/Southern Oscillation. *Max-Planck-Institut für Meteorologie*, Hamburg, Report **129**.

LORENZ, S., GRIEGER, B., HELBIG, P., HERTERICH, K., 1996. Investigating the sensitivity of the atmospheric general circulation model ECHAM3 to palaeoclimatic boundary conditions. *Geologische Rundschau*, **85(3)**, 513-524.

LUNKEIT, F., FRAEDRICH, K. BAUER, S.E., 1998. Storm tracks in a warmer climate: sensitivity studies with a simplified global circulation model. *Climate Dynamics*, **14**, 813-826.

MACFADDEN, B.J., WANG, Y., CERLING, T.E., ANAYA, F., 1994. South American fossil mammals and carbon isotopes: a 25 million year sequence from the Bolivian Andes. *Palaeogeography, Palaeoclimatology, Palaeoecology*, **107**, 257-268.

MAI, D.H., 1995. Tertiäre Vegetationsgeschichte Europas. *Gustav Fischer*, Jena.

MAIER-REIMER, E., MIKOLAJEWICZ, U., CROWLEY, T.J., 1990. Ocean general circulation model sensitivity experiment with an open Central American Isthmus. *Palaeoceanography*, **5(3)**, 349-366.

MANABE, S., STOUFFER, R.J., 1997. Coupled ocean-atmosphere model response to freshwater input: comparison to Younger Dryas event. *Palaeoceanography*, **12**, 321-336.

MARTIN, H.A., 1990. Tertiary climate and phytogeography in southeastern Australia. *Review of Palaeobotany and Palynology*, **65**, 47-55.

MARTIN, H.A., 1998. Tertiary climatic evolution and the development of aridity in Australia. *Proc. Linn. Soc. N.S.W.*, **119**, 115-136.

MATTHEWS, E., 1983. Global vegetation and land use: new high resolution data bases for climate studies. *J. Clim. Appl. Meteor.*, **22**, 474-487.

MELLILO, J., MCGUIRE, A., KICKLIGHTER, D., MOORE III, B., VOROSMARTY, C., SCHLOSS, A., 1993. Global climate change and terrestrial net primary production. *Nature*, **363**, 234-240.

May, G., Hartley, A.J., Stuart, F.M., Chong, G., 1999. Tectonic signatures in arid continental basins: an example from the Upper Miocene-Pleistocene, Calama basin, Andean forearc, northern Chile. *Palaeogeography, Palaeoclimatology, Palaeoecology*, **151**, 55-77.

McCartan, L., Tiffney, B.H., Wolfe, J.A., Ager, T.A., Wing, S.L., Sirkin, L.A., Ward, L.W., Brooks, J., 1990. Late Tertiary floral assemblage from upland gravel deposits of the southern Maryland Coastal Plain. *Geology*, **18**, 311-314.

McGuffie, K., Henderson-Sellers, A., Holbrook, N., Kothavala, Z., Balcjova, O., Hoekstra, J., 1999. Assessing simulations of daily temperatures and precipitation variability with global climate models for present and enhanced greenhouse climates. *International Journal of Climatology*, **19**, 1-26.

Mikolajewicz, U., Maier-Reimer, E., Crowley, T.J., Kim, K.J., 1993. Effect of Drake and Panamanian gateways on the circulation of an ocean model. *Palaeoceanography*, **8(4)**, 409-426.

Mikolajewicz, U., Crowley, T.J., 1997. Response of a coupled ocean/energy balance model to restricted flow through the central American isthmus. *Palaeoceanography*, **12(3)**, 429-441.

Mikolajewicz, U., Voss, R., 2000. The role of the individual air-sea flux components in CO_2-induced changes of the ocean's circulation and climate. *Climate Dynamics*, **16**, 627-642.

Montoya, M., Crowley, T.J., von Storch, H., 1998. Temperatures at the last interglacial simulated by a coupled ocean-atmosphere climate model. *Paleoceanography*, **13(2)**, 170-177.

Mosbrugger, V., Schilling, H.-D., 1992. Terrestrial palaeoclimatology in the Tertiary: a methodological critique. *Palaeogeography, Palaeoclimatology, Palaeoecology*, **99**, 17-29.

Mosbrugger, V., Utescher, T., 1997. The coexistence approach - a method for quantitative reconstructions of Tertiary terrestrial palaeoclimate data using plant fossils. *Palaeogeography, Palaeoclimatology, Palaeoecology*, **134**, 61-86.

Mudie, P.J., Helgason, J., 1983. Palynological evidence for Miocene climatic cooling in eastern Iceland 9.8 Myr ago. *Nature*, **303**, 689-692.

Nemry, B., Francois, L.M., Warnant, P., Robinet, F., Gerard, J.-C., 1996. The seasonality of the CO_2 exchange between the atmosphere and the land biosphere: A study with a global mechanistic vegetation model. *Journal of Geophysical Research*, **101(D3)**, 7111-7125.

New, M., Hulme, M., Jones, P., 1999. Representing Twentieth-Century Space-Time Climate Variability. Part I: Development of a 1961-90 Mean Monthly Terrestrial Climatology. *Journal of Climate*, **12**, 829-856.

Otto, D., Rasse, D., Kaplan, J., Warnant, P., Francois, L., 2002. Biospheric carbon stocks reconstructed at the Last Glacial Maximum: comparison between general circulation models using prescribed and computed sea surface temperatures. *Global and Planetary Change*, **33**, 117-138.

Otto-Bliesner, B., Upchurch, G.R. Jr., 1997. Vegetation-induced warming of high-latitude regions during the Late Cretaceous period. *Nature*, **385**, 804-807.

PAGANI, M., ARTHUR, M.A., FREEMAN, K.H., 1999. Miocene evolution of atmospheric carbon dioxide. *Paleoceanography*, **14(3)**, 273-292.

PARTRIDGE, T.C., BOND, G.C., HARTNADY, C.J.H., DEMENOCAL, P.B., RUDDIMAN, W.F., 1995. Climatic effects of the Late Neogene tectonism and volcanism. In: VRBA, E.S., G.H. DENTON, T.C. PARTRIDGE, L.H. BURCKLE (Eds.). Paleoclimate and evolution with emphasis on human origins. *Yale University Press*, London, 547pp.

PEARSON, P.N., PALMER, M.R., 2000. Atmospheric carbon dioxide concentrations over the past 60 million years. *Nature*, **406**, 695-699.

PEARSON, P.N., DITCHFIELD, P.W., SINGANO, J., HARCOURT-BROWN, K.G., NICHOLAS, C.J., OLSSON, R.K., SHACKLETON, N.J., HALL, M.A., 2001. Warm tropical sea surface temperatures in the Late Cretaceous and Eocene epochs. *Nature*, **413**, 481-487.

PLAYFORD, G., 1982. Neogene palynomorphs from the Huon Penninsula, Papua New Guinea. *Palynology*, **6**, 29-54.

PRELL, W.L., KUTZBACH, J.E., 1992. Sensitivity of the Indian monsoon to forcing parameters and implications for its evolution. *Nature*, **360**, 647-652.

PRENTICE, K.C., FUNG, I.Y., 1990. The sensitivity of terrestrial carbon storage to climate change. *Nature*, **346**, 48-51.

PRENTICE, C., CRAMER, W., HARRISON, S.P., LEEMANS, R., MONSERUD, R.A., SOLOMON, A.M., 1992. A global biome model based on plant physiology and dominance, soil properties and climate. *Journal Of Biogeography*, **19**, 117-134.

RAHMSTORF, S., 1995. Bifurcation of the Atlantic thermohaline circulation in response to changes in the hydrological cycle. *Nature*, **378**, 145-149.

RAMSTEIN, G., FLUTEAU, F., BESSE, J., JOUSSAUME, S., 1997. Effect of orogeny, plate motion and land-sea distribution on Eurasian climate change over past 30 million years. *Nature*, **386**, 788-795.

RAYMO, M.E., RIND, D., RUDDIMAN, W.F., 1990. Climate effects of reduced arctic sea ice limits in GISS II general circulation model. *Paleoceanography*, **5(3)**, 367-382.

ROECKNER, E., ARPE, K., BENGTSSON, L., BRINKOP, S., DÜMENIL, L., ESCH, M., KIRK, E., LUNKEIT, F., PONATER, M., ROCKEL, B., SAUSEN, R., SCHLESE, U., SCHUBERT, S., WINDELBAND, M., 1992. Simulation of the present-day climate with the ECHAM model: impact of model physics and resolution. *Max-Planck-Institut für Meteorologie*, Hamburg, Report **93**.

ROECKNER, E., ARPE, K., BENGTSSON, L., CHRISTOPH, M., CLAUSSEN, M., DÜMENIL, L., ESCH, M., GIORGETTA, M., SCHLESE, U., SCHULZWEIDA, U., 1996. The atmospheric general circulation model ECHAM-4: Model description and simulation of present-day climate. *Max-Planck-Institut für Meteorologie*, Hamburg, Report **218**.

ROGERS, J., 1997. North Atlantic storm track variability and its association to the North Atlantic Oscillation and climate variability of Northern Europe. *Journal of Climate*, **10**, 1635-1647.

References

RUDDIMAN, W.F., KUTZBACH, J.E., 1989. Forcing of late Cenozoic northern hemisphere climate by uplift in southern Asia and the American West. *Journal of Geophysical Research*, **94**, 18409-18427.

SACHSE, M., MOHR, B.A.R., 1996. Eine obermiozäne Makro- und Mikroflora aus Südkreta (Griechenland), und deren paläoklimatische Interpretation. - Vorläufige Betrachtungen. *N. Jb. Geol. Paläont. Abh.*, **200(1/2)**, 149-182.

SACHSE, M., unpublished. Die Makrilia-Flora (Kreta, Griechenland) - Ein Beitrag zur Neogenen Klima- und Vegetationsgeschichte des östlichen Mittelmeergebietes. *Unpublished Doctoral thesis*, ETH Zürich, Nr. **12250**.

SAKAI, H., 1997. When monsoon climate was set up? Its geological evidence. *Journal of Geography*. **106**, 131-144.

SCHAEFFER, R., SPIEGLER, D., 1986. Neogene Kälteeinbrüche und Vereisungsphasen im Nordatlantik. *Z. Dtsch. Geol. Ges.*, **137**, 537-552.

SCHNITZLER, K.-G., KNORR, W., LATIF, M., BADER, J., ZENG, N., 2001. Vegetation feedback on Sahelian rainfall and variability in a coupled climate land – vegetation model. *Max-Planck-Institut für Meteorologie*, Hamburg, Report **329**.

SCHUBERT, M., PERLWITZ, J., BLENDER, R., FRAEDRICH, K., LUNKEIT, F., 1998. North Atlantic cyclones in CO_2-induced warm climate simulations: frequency, intensity, and tracks. *Climate Dynamics*, **14**, 827-837.

SIMMONS, A.J., BRRIDGE, D.M., JARRAUD, M., GIRARD, C., WERGEN, W., 1989. The ECMWF medium-range prediction models: Development of the numerical formulations and the impact of increased resolution. *Meteorol. Atmos. Phys.*, **40**, 28-60.

STEPPUHN, A., 2002. Climate and climate processes during the Upper Miocene: Sensitivity studies with coupled general circulation models. *Doctoral thesis*, University of Tübingen, http://w210.ub.uni-tuebingen.de/dbt/volltexte/2002/533/.

STEPPUHN, A., MICHEELS, A., GEIGER, G., MOSBRUGGER, V., 2006. Reconstructing the Late Miocene climate and oceanic heat flux using the AGCM ECHAM4 coupled to a mixed layer ocean model with adjusted flux correction. *Palaeogeography, Palaeoclimatology, Palaeoecology*, **238**, 399-423.

STEPPUHN, A., MICHEELS, A., BRUCH, A.A., UTESCHER, T., MOSBRUGGER, V., 2007. The sensitivity of ECHAM4/ML to a double CO_2 scenario for the Late Miocene and the comparison to terrestrial proxy data, *Global and Planetary Change*, **57**, 189–212.

STRÖMBERG, C.A.E., 2002. The origin and spread of grass-dominated ecosystems in the late Tertiary of North America: preliminary results concerning the evolution of hypsodonty. *Palaeogeography, Paleoclimatology, Palaeoecology*, **177**, 59-75.

THIEDE, J., WINKLER, A., WOLF-WELLING, T., ELDHOLM, O., MYHRE, A.M., BAUMANN, K.-H., HENRICH, R., STEIN, R., 1998. Late Cenozoic history of the Polar North Atlantic: results from ocean drilling. *Quaternary Science*, **17**, 185-208.

THONICKE, K., VENEVSKY, S., SITCH, S., CRAMER, W., 2001. The role of fire disturbance for global vegetation dynamics: coupling fire into Dynamic Global Vegetation Model. *Global Ecology and Biogeography*, **10**, 661-677.

TRAISER, C., submitted. Environmental signals from leaves - Analysis of extant and fossil leaf assemblages. *Doctoral thesis*, University of Tübingen.

TSUCHI, R., 1997. Marine climatic responses to Neogene tectonics of the pacific ocean seaways. *Tectonophysics*, **281**, 113.124.

UNGER, H.J., 1983. Die Makro-Flora der Mergelgrube Aubenham nebst Bemerkungen zur Lithologie, Ökologie und Stratigraphie. *Geol. Jahrbuch*, **67**, 37-129.

UPCHURCH, G.R. JR., OTTO-BLIESNER, B.L., SCOTESE, C.R., 1998. Vegetation-atmosphere interactions and their role in global warming during the latest Cretaceous. *Philosophical transactions of the Royal Society of London/B.*, **353**, 97-112.

UPCHURCH, G.R. JR., OTTO-BLIESNER, B.L., SCOTESE, C.R., 1999. Terrestrial vegetation and its effect on climate during the latest Cretaceous. *Geological Society of America*, Special Paper **332**, 407-426.

UTESCHER, T., GEBKA, M., MOSBRUGGER, V., SCHILLING, H.-D., ASHRAF, A.R., 1997. Regional palaeontological-meteorological palaeoclimate reconstruction of the Neogene Lower Rhine Embayment. *Proceedings 4th, EPPC*, **58**, 263-271.

UTESCHER, T., MOSBRUGGER, V., ASHRAF, A.R., 2000. Terrestrial climate evolution in Northwest Germany over the last 25 million years. *Palaios*, **15(5)**, 430-449.

VAN DAM, J.A., WELTJE, G.J., 1999. Reconstruction of the Late Miocene climate of Spain using rodent palaeocommunity successions: an application of end-member modelling. *Palaeogeography, Palaeoclimatology, Palaeoecology*, **151**, 267-305.

VAN DER BURGH, J., VISSCHER, H., DILCHER, D.L., KÜRSCHNER, W.M., 1993. Paleoatmospheric Signatures in Neogene Fossil Leaves. *Science*, **260**, 1788-1790.

VON STORCH, H., ZWIERS, F.W., 1999. Statistical Analysis in Climate Research. *Cambridge University Press*, Cambridge, 484pp.

WANG, J., WANG, Y.J., LIU, Z.C., LI, J.Q., XI, P., 1999. Cenozoic environmental evolution of the Qaidam Basin and its implications for the uplift of the Tibetan Plateau and the drying of central Asia. *Palaeogeography, Palaeoclimatology, Palaeoecology*, **152**, 37-47.

WARNANT, P., FRANCOIS, L.M., STRIVAY, D., GERARD, J.-C., 1994. CARAIB: A global model of terrestrial biological productivity. *Global Biogeochemical Cycles*, **8(3)**, 255-270.

WEBSTER, P.J., PALMER, T.N., 1997. The past and the future of El Niño. *Nature*, **390**, 562-564.

WERNER, M., MIKOLAJEWICZ, U., HOFFMANN, G., HEIMANN, M., 1999. Possible changes of $\delta^{18}O$ in precipitation caused by a meltwater event in the North Atlantic. *Max-Planck-Institut für Meteorologie*, Hamburg, Report **294**.

WHITE, J.M., AGER, T.A., ADAM, D.P., LEOPOLD, E.B., LIU, G., JETTE, H., SCHWEGER, C.E., 1997. An 18 million year record of vegetation and climate change in northwestern Canada and Alaska: tectonic and global climatic correlates. *Palaeogeography, Palaeoclimatology, Palaeoecology*, **130**, 293-306.

WOLF, T.C.W., THIEDE, J., 1991. History of terrigenous sedimentation during the past 10 m.y. in the North Atlantic (ODP Legs 104 and 105 and DSDP Leg 81). *Marine Geology*, **101**, 83-102.

WOLFE, J.A., 1985. Distribution of major vegetational types during the Tertiary. In: The carbon cycle and atmospheric CO_2 Natural variations Archean to Present [SUNDQUIST, E.T., BROECKER, W.S. (eds.)]. *American Geophysical Union*, Washington D.C., 357-375.

WOLFE, J.A., 1993. A method of obtaining climatic parameters from leaf assemblages. *U.S. Geological Survey Bulletin*, **2040**, 73pp.

WOLFE, J.A., 1994a. Tertiary climatic changes at middle latitudes of western North America. *Palaeogeography, Palaeoclimatology, Palaeoecology*, **108**, 195-206.

WOLFE, J.A., 1994b. An analysis of Neogene climates in Beringia. *Palaeogeography, Palaeoclimatology, Palaeoecology*, **108**, 207-216.

WOODRUFF, F., SAVIN, S.M., 1989. Miocene deepwater oceanography. *Palaeoceanography*, **4**, 87-140.

WU, X., WANG, S., AN, Z., JIANG, F. XIAO, H., SUN, D., XUE, B., 1998. On tectonoclimatic cycle of quasi-period of 1.2 Ma in the late Cenozoic; examples from Qinghai-Xizang Plateau and Loess Plateau, China. *Journal of Geomechanics*, **4**, 1-11.

APPENDIX A - THE PALVEG TORTONIAN RUN WITH ECHAM4/ML

To figure out the effects of the palaeovegetation and to see the difference between the Tortonian and the present-day's climate, differences plots (the PalVeg Tortonian run minus the Standard Tortonian and the PalVeg Tortonian run minus the Recent Control) were used in sec.4. Figures A1 to A11 show selected absolute data fields of the PalVeg Tortonian run.

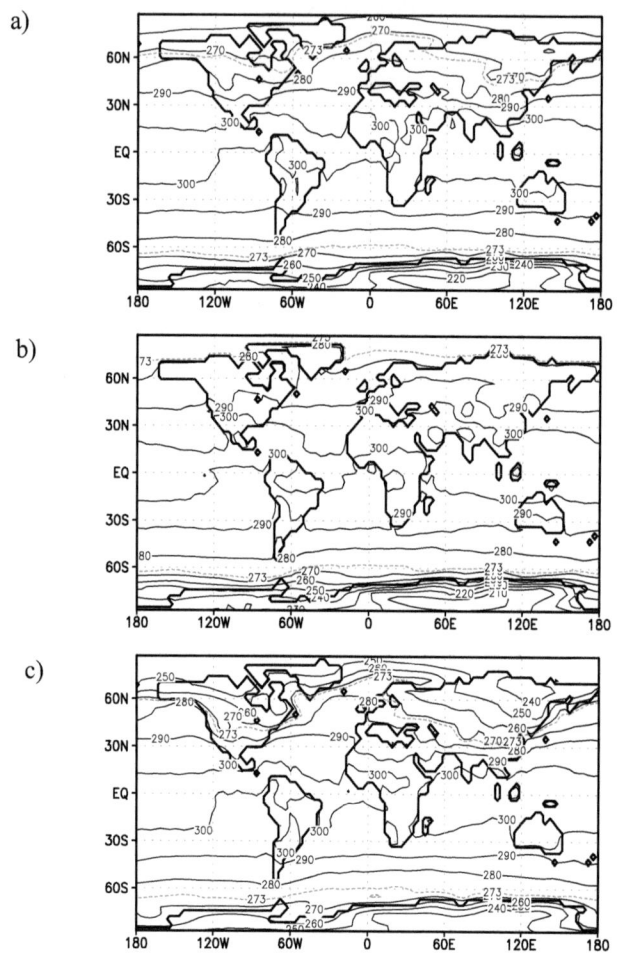

Figure A.1: *The 2m-temperature of the PalVeg Tortonian run [K] for a) the annual average, b) JJA and c) DJF. The contour intervals are 10 K and the 273 K-isotherme is shown grey dotted.*

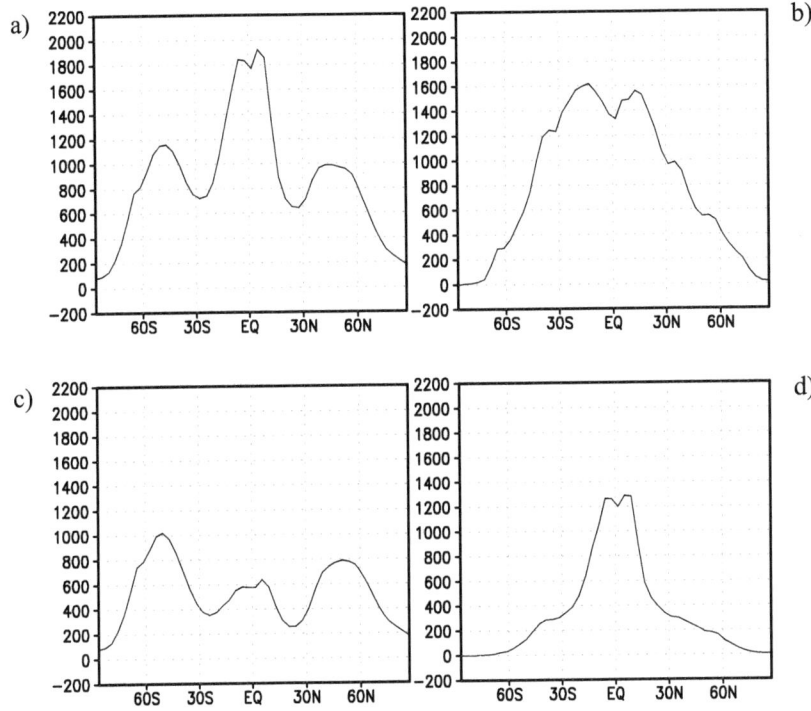

Figure A.2: *The annual and zonal averages of the PalVeg Tortonian run for a) the total precipitation rate [mm/a], b) the evaporation rate [mm/a], c) the large-scale precipitation rate [mm/a] and d) the convective precipitation rate [mm/a].*

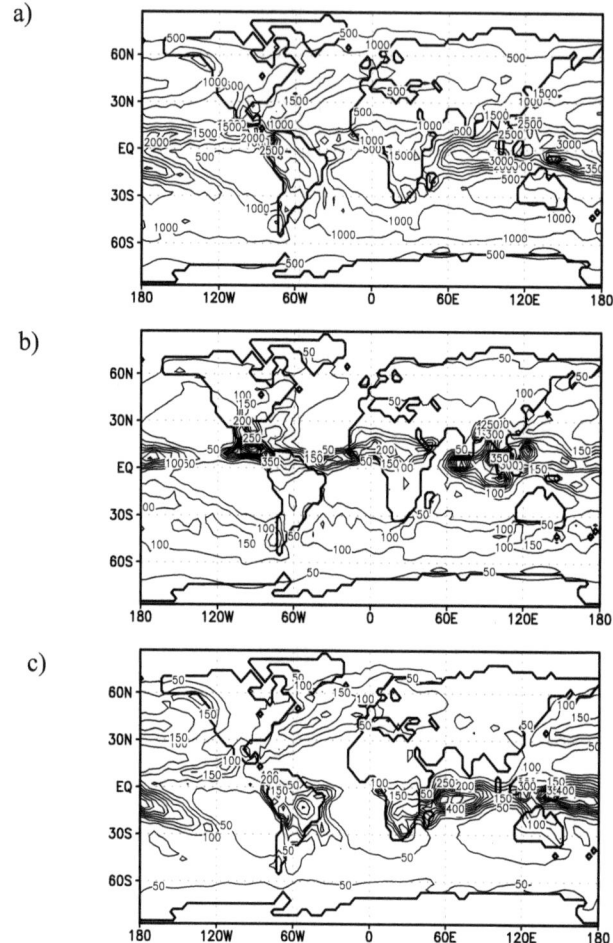

Figure A.3: *The total precipitation rate of the PalVeg Tortonian run for a) the annual average [mm/a], b) JJA and [mm/mon] c) DJF [mm/mon]. The contour intervals are 500 mm/a for a) and 50 mm/mon for b) and c).*

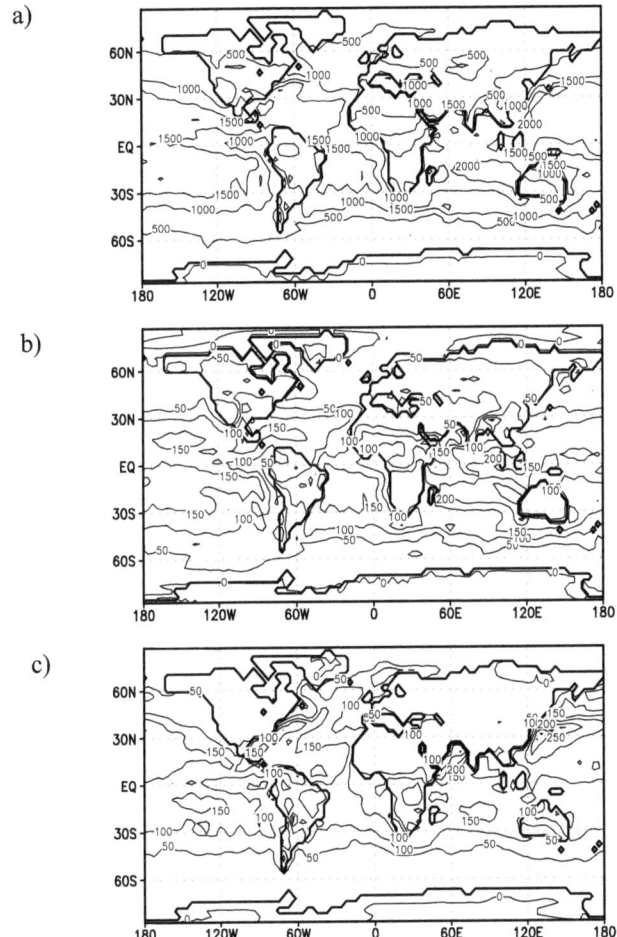

Figure A.4: *The evaporation rate of the PalVeg Tortonian run for a) the annual average [mm/a], b) JJA and [mm/mon] c) DJF [mm/mon]. The contour intervals are 500 mm/a for a) and 50 mm/mon for b) and c).*

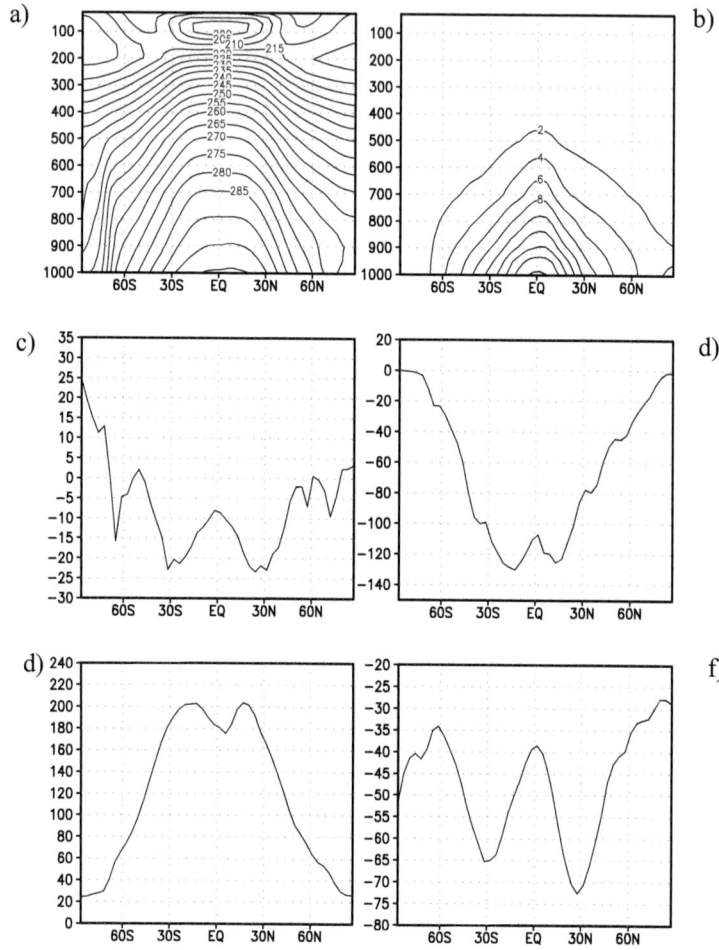

Figure A.5: *The annual and zonal averages of a) the temperature [K], b) the specific humidity [g/kg], c) the surface sensible heat flux [W/m²], d) the surface latent heat flux [W/m²], e) the net surface solar radiation flux [W/m²] and f) the surface terrestrial radiation flux [W/m²] of the PalVeg Tortonian run. a) and b) are shown with respect to height (in pressure coordinates). The contour intervals are 5 K in a) and 2 g/kg in b).*

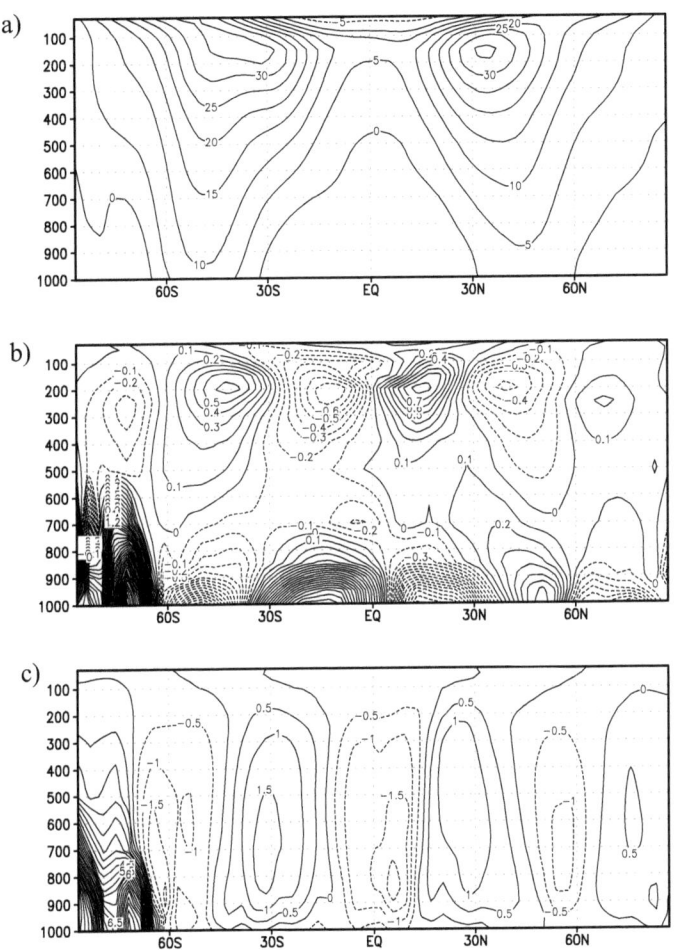

Figure A.6: *The annual and zonal averages of a) the zonal wind [m/s], b) the meridional wind [m/s] and c) the vertical wind [Pa/s] with respect to height (in pressure coordinates [hPa]) of the PalVeg Tortonian run. The contour intervals are 5 m/s for a), 0.1 m/s for b) and 0.5×10^{-2} Pa/s for c).*

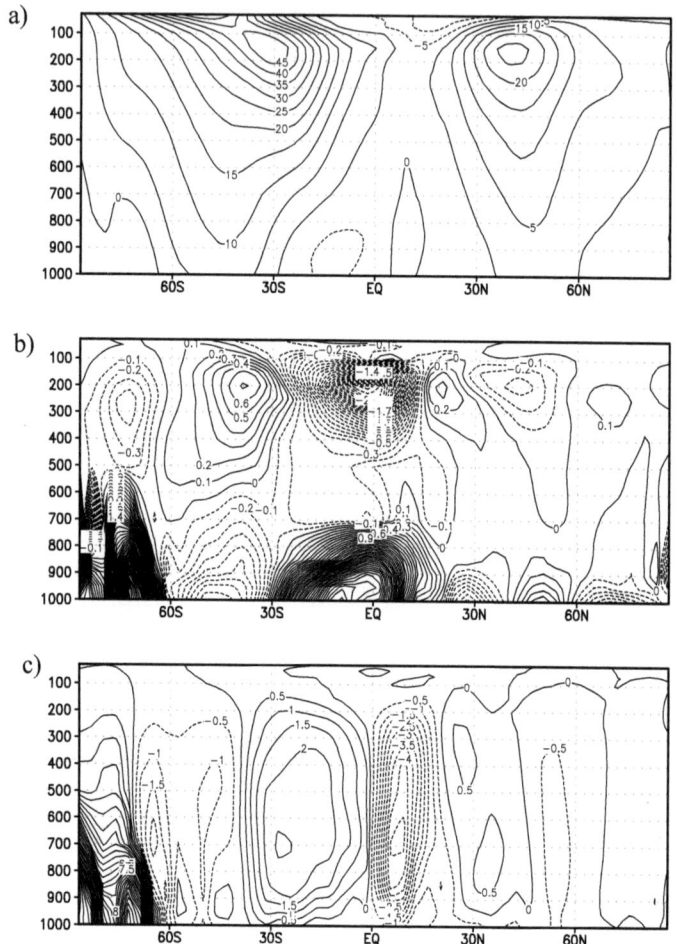

Figure A.7: *The same as in fig.A.6 but for JJA.*

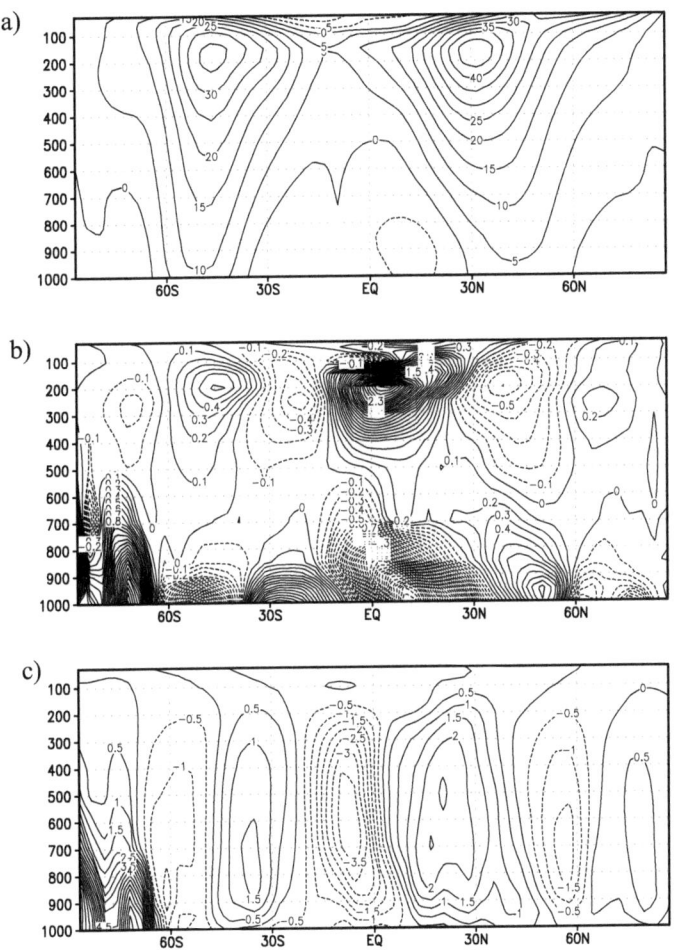

Figure A.8: *The same as in fig.A.6 but for DJF.*

APPENDIX A - THE PALVEG TORTONIAN RUN WITH ECHAM4/ML

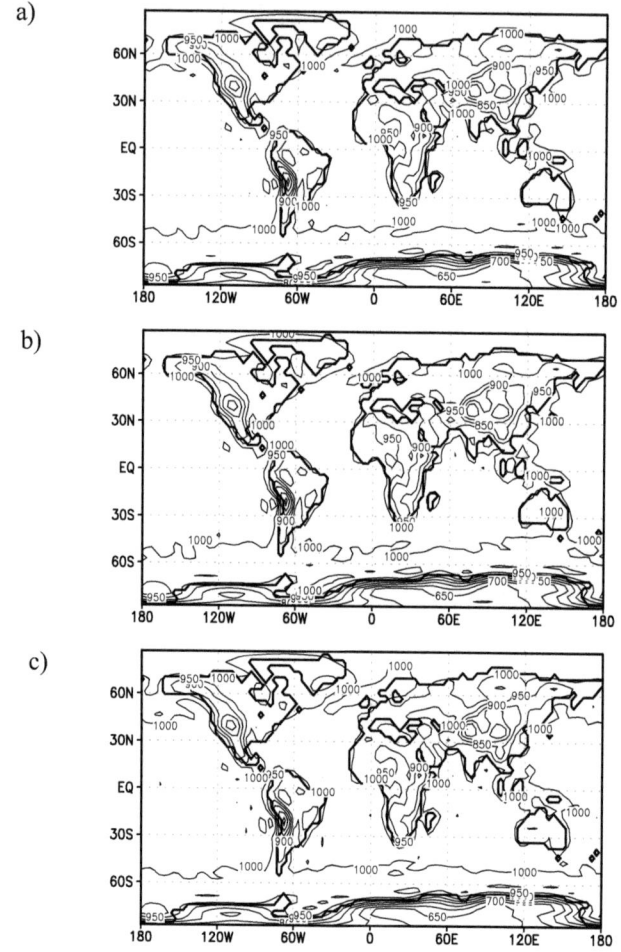

Figure A.9: *The surface pressure of the PalVeg Tortonian run [hPa] for a) the annual average, b) JJA and c) DJF. The contour intervals are 50 hPa.*

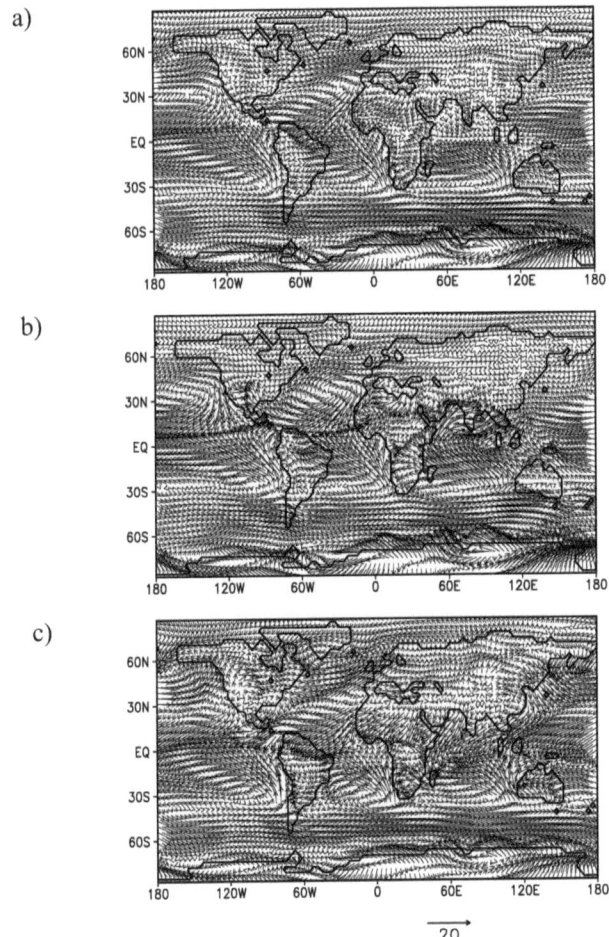

Figure A.10: *The horizontal wind at 1000 hPa of the PalVeg Tortonian run [m/s] for a) the annual average, b) JJA and c) DJF. The reference arrow is 20 m/s.*

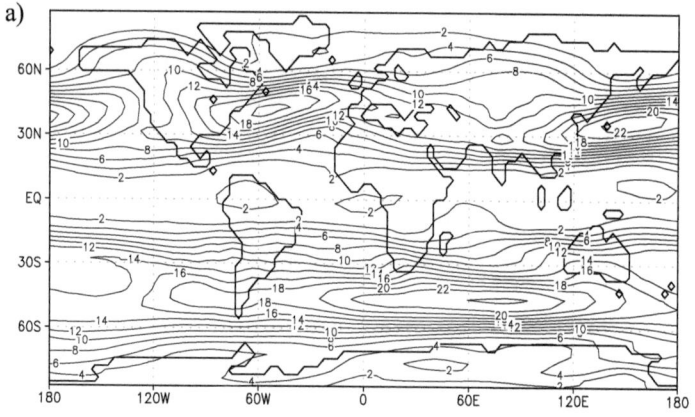

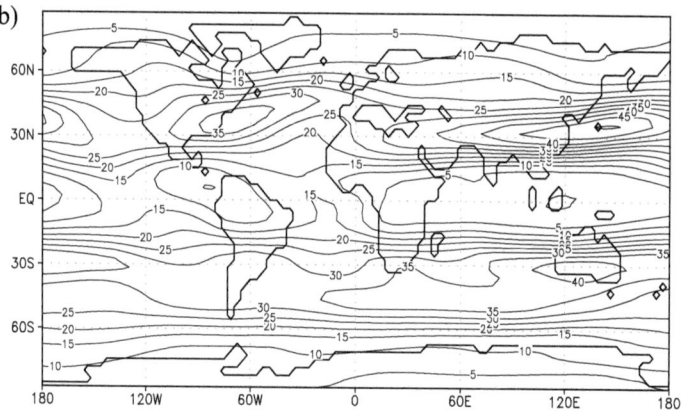

Figure A.11: *The annual average wind speed of the PalVeg Tortonian run [m/s] at a) 500 hPa and b) 200 hPa. The contour intervals are 2 m/s in a) and 5 m/s in b).*

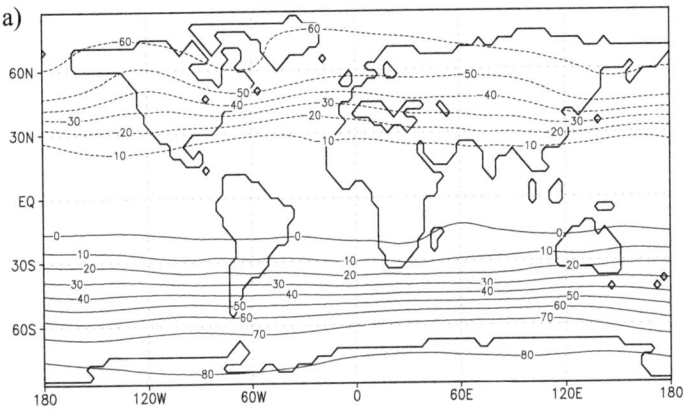

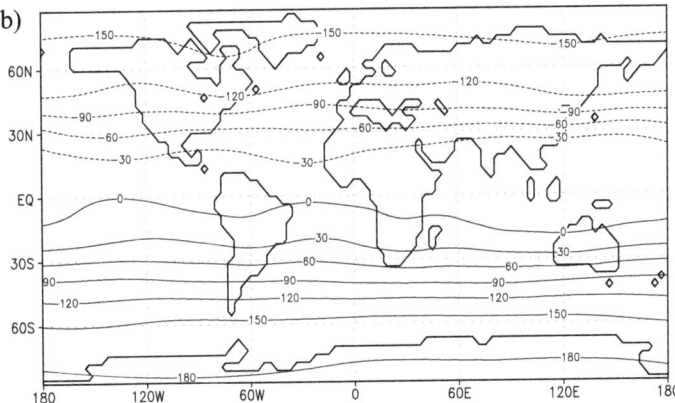

Figure A.12: *The annual average streamfunction of the PalVeg Tortonian run [10^6 m^2/s] at a) 500 hPa and b) 200 hPa. The contour intervals are 10×10^6 m^2/s in a) and 30×10^6 m^2/s in b).*

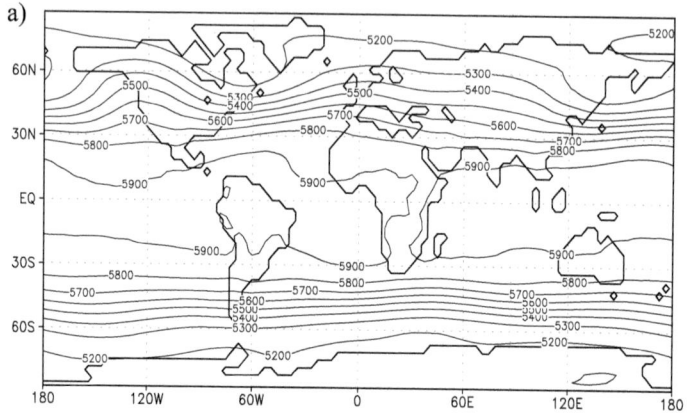

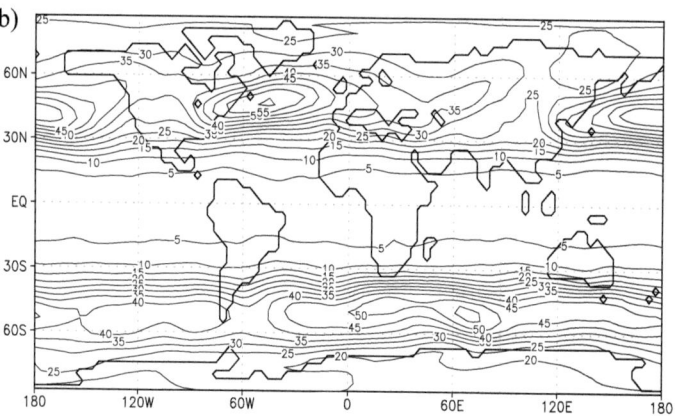

Figure A.13: a) The 500 hPa-geopotential field of the PalVeg Tortonian run [gpm] (DJF) and b) the band-pass filtered standard deviation of the 500 hPa-geopotential field of the PalVeg Tortonian run [gpm] (DJF). The contour intervals are 100 gpm in a) and 5 gpm in b).

Appendix B - List of Symbols

α_v	[-]	albedo of vegetation
α_{min}	[-]	minimum moisture index
α_{max}	[-]	maximum moisture index
c_{veg}	[-]	fractional vegetation cover
c_{for}	[-]	fractional forest cover
Δ		denotes differences of any variables
D	[-]	dominance hierarchy
δ		denotes uncertainties of any variables
$\delta^{18}O$	[%]	oxygen isotope ratio
DJF		months December, January and February
E	[mm/a],[mm/mon]	evaporation rate
gdd0	[°C]	temperature sum of days above 0°C
gdd5	[°C]	temperature sum of days above 5°C
JJA		months June, July and August
LAI	[m²/m²]	leaf area index
MAT	[°C], [K]	mean annual temperature
MAP	[mm/a]	mean annual precipitation rate
NAO		North Atlantic Oscillation index
p	[-]	level of significance used for the Student t-test
p_{tot}	[mm/a], [mm/mon]	total precipitation rate (= large-scale + convective precipitation rate)
PFT		plant functional type
q		climatological field variable
q_{obs}		observed climatological field variable
SST	[°C], [K]	sea surface temperature
T	[°C], [K]	temperature
T_{2m}	[°C], [K]	temperature at 2m height
$T_{c,min}$	[°C], [K]	minimum temperature of the coldest month
$T_{c,max}$	[°C], [K]	maximum temperature of the coldest month
T_{global}	[°C], [K]	global average temperature
T_{NH}	[°C], [K]	average temperature of the Northern Hemisphere
T_{SH}	[°C], [K]	average temperature of the Southern Hemisphere
T_s	[°C], [K]	surface temperature
$T_{w,min}$	[°C], [K]	minimum temperature of the warmest month
$W_{s,max}$	[m]	maximum available soil water capacity
z_0	[m]	surface roughness length
$z_{0,oro}$	[m]	orography roughness length
$z_{0,veg}$	[m]	vegetation roughness length

Die VDM Verlagsservicegesellschaft sucht für wissenschaftliche Verlage abgeschlossene und herausragende

Dissertationen, Habilitationen, Diplomarbeiten, Master Theses, Magisterarbeiten usw.

für die kostenlose Publikation als Fachbuch.

Sie verfügen über eine Arbeit, die hohen inhaltlichen und formalen Ansprüchen genügt, und haben Interesse an einer honorarvergüteten Publikation?

Dann senden Sie bitte erste Informationen über sich und Ihre Arbeit per Email an *info@vdm-vsg.de*.

Sie erhalten kurzfristig unser Feedback!

VDM Verlagsservicegesellschaft mbH
Dudweiler Landstr. 99
D - 66123 Saarbrücken
www.vdm-vsg.de

Telefon +49 681 3720 174
Fax   +49 681 3720 1749

Die VDM Verlagsservicegesellschaft mbH vertritt

Printed by Books on Demand GmbH, Norderstedt / Germany